Complex Orthogonal Space-Time Processing In Wireless Communications

COMPLEX ORTHOGONAL SPACE-TIME PROCESSING IN WIRELESS COMMUNICATIONS

LE CHUNG TRAN
School of Elec., Comp. and Telecom. Eng.
University of Wollongong
Northfields Avenue, Wollongong, NSW 2522,
Australia

TADEUSZ A. WYSOCKI
School of Elec., Comp. and Telecom. Eng.
University of Wollongong
Northfields Avenue, Wollongong, NSW 2522,
Australia

ALFRED MERTINS
Department of Physics
University of Oldenburg
26111 Oldenburg,
Germany

JENNIFER SEBERRY
School of Inform. Technol. and Comp. Science
University of Wollongong
Northfields Avenue, Wollongong, NSW 2522,
Australia

Springer

LE CHUNG TRAN
TADEUSZ A. WYSOCKI
School of Elec., Comp. and Telecom. Eng.
University of Wollongong
Northfields Avenue, Wollongong, NSW 2522
Australia

ALFRED MERTINS
Dept. of Physics
University of Oldenburg
26111 Oldenburg
Germany

JENNIFER SEBERRY
School of Inform. Tech. and Comp. Science
University of Wollongong
Northfields Avenue, Wollongong, NSW 2522
Australia

Complex Orthogonal Space-Time Processing in Wireless Communications

Library of Congress Control Number: 2005933881

ISBN 0-387-29291-8   e-ISBN 0-387-29544-5
ISBN 978-0387-29291-5

Printed on acid-free paper.

Printed in the United States of America.

9 8 7 6 5 4 3 2 1   SPIN 11556640

springeronline.com

This book is dedicated to our families for their enormous support and encouragement.

Contents

List of Figures

List of Tables

Foreword

Multiple-Input Multiple-Output (MIMO) systems have recently been the subject of intensive consideration in modern wireless communications as they offer the potential of providing high capacity, thus unleashing a wide range of applications in the wireless domain. The main feature of MIMO systems is the use of space-time processing and Space-Time Codes (STCs). Among a variety of STCs, orthogonal Space-Time Block Codes (STBCs) have a much simpler decoding method, compared to other STCs.

This book provides an in depth understanding of space-time processing in general and space-time block processing in particular including their applications in MIMO wireless communication systems. Importantly, the book provides readers, for the first time, with state of the art critical reviews and findings in the area of space-time processing. The authors' latest discoveries in the field of complex orthogonal space-time processing for wireless communications represent the core contributions of the book along an overview of open research issues.

This book is considered suitable for both general and professional audiences in the areas of Communications, Vehicular Technology, Signal Processing, and to some extent Information Theory. In many ways, it can be considered as a supplementary text for the standard courses in advanced wireless communications as well as a fundamental source of knowledge for future research in the area of orthogonal space-time processing.

One anticipates that this book will provide lasting values for both research and educational purposes.

Professor Joe Chicharo, PhD, FIE (Australia), SMIEEE
Dean of Faculty of Informatics
University of Wollongong, Australia

Preface

Digital communication using Multiple-Input Multiple-Output (MIMO) systems has recently emerged as one of the most significant technical breakthroughs in modern communication. Communication theories show that MIMO systems can potentially provide potentially a very high capacity that, in many cases, grows approximately linear with the number of antennas. Therefore, MIMO transmission is an outstanding technique with a chance to resolve the bottleneck of traffic capacity in the future wireless networks.

The main feature of MIMO systems is *space-time* processing. Space-Time Codes (STCs) are the codes designed for the use in MIMO systems. Among a variety of STCs, *orthogonal* Space-Time Block Codes (STBCs) possess a much simpler decoding method over other STCs. Because of that, this book examines *orthogonal* STBCs in MIMO systems. Furthermore, Complex Orthogonal STBCs (CO STBCs) are *mainly* considered in this book since they can be used for PSK/QAM modulation schemes, and therefore, are more practical than real STBCs.

The book starts with the backgrounds on MIMO systems and their capacity, on STBCs, and on some conventional transmission diversity techniques. After reviewing the state of the art of the issues related to this book and indicating the gaps in the literature, we mention the following topics:

Novel constructions methods for improved, square CO STBCs

In this book, we first propose three new, *maximum rate*, order-8 CO STBCs. These new CO STBCs are amenable to practical implementations as they allow for a more uniform spread of power among the transmitter antennas, while providing better performance than other published codes of order 8 for the same peak power per transmitter antenna.

Constructions of *square, maximum rate* CO STBCs are well known, however codes constructed via the known methods include numerous zeros, which impede their practical implementation, especially in high data rate systems. This disadvantage is partially overcome by the three new CO STBCs of order 8 mentioned above. However, these new codes still contain zeros which are undesirable or the design method is *neither* general *nor* easy yet.

Hence, later, we discover two construction methods of *square*, order-$4n$ CO STBCs from *square*, order-n codes which satisfy certain properties, by modifying the Williamson and the Wallis-Whiteman arrays to apply to complex matrices. Applying the proposed methods, we construct *square, maximum rate*, order-8 CO STBCs with no zeros, such that the transmitted symbols equally disperse through transmitter antennas. Those codes, referred to as the *improved square CO STBCs*, have the advantages that the power is equally transmitted via each transmitter antenna during every symbol time slot and that a lower peak power per transmitter antenna is required to achieve the same bit error rates as in the conventional CO STBCs with zeros.

Multi-modulation schemes to increase the data rate of CO STBCs

Based on the new proposed CO STBCs, multi-modulation schemes (MMSs) are proposed to increase the information transmission rate of those new codes of order 8. Simulation results show that, for the same MMSs and the same peak power per transmitter antenna, the three new codes provide better error performance than the conventional CO STBCs of the same order 8.

In addition, the method to evaluate the optimal inter-symbol power allocation in the proposed codes in single modulation as well as in different MMSs for both Additive White Gaussian Noise (AWGN) and flat Rayleigh fading channels is derived. It turns out that, for some modulation schemes, equal power transmission per symbol time slot is not only optimal from the technical point of view, but also optimal in terms of achieving the best symbol error probability.

The MMSs which increase the information transmission rate of CO STBCs and the method to examine the optimal power allocation for multi-modulated CO STBCs mentioned here can be generalized for CO STBCs of other orders without any difficulty.

Transmitter diversity antenna selection techniques for MIMO systems using STBCs and DSTBCs

The combination of CO STBCs and a closed loop transmission diversity technique using a feedback loop has received a considerable attention in the literature since it allows us to improve performance of wireless communication channels with coherent detection. In this book, we propose an improved diversity Antenna Selection Technique (AST) to improve further the performance of such channels. Calculations and simulations show that our technique performs well, especially, when it is combined with the Alamouti code.

While the combination between STBCs and a closed loop transmission diversity technique in the case of *coherent detection* has been intensively considered in the literature, it seems to be missing for the case of *differential detection*. The book thus proposes two ASTs for wireless channels utilizing Differential Space-Time Block Codes (DSTBCs), which are referred to as the AST/DSTBC schemes. These techniques remarkably improve the performance of wireless channels using DSTBCs (with *differential detection*).

Effects of imperfect channels on transmitter diversity antenna selection techniques

The proposed AST/DSTBC schemes work very well in independent, flat Rayleigh fading channels as well as in the case of perfect carrier recovery. How do they perform in the case of correlated, flat Rayleigh fading channels or in the case of imperfect carrier recovery?

To answer this question, first, we propose here a very general, straightforward algorithm for the generation of an *arbitrary number* of Rayleigh envelopes with any desired (*equal* or *unequal*) power, in wireless channels either *with* or *without* Doppler frequency shift effects. The proposed algorithm can be applied to the case of *spatial correlation*, such as with antenna arrays in Multiple Input Multiple Output (MIMO) systems, or *spectral correlation* between the random processes like in Orthogonal Frequency Division Multiplexing (OFDM) systems. It can also be used for generating correlated Rayleigh fading envelopes in either *discrete-time instants* or a *real-time scenario*. Besides being *more generalized*, our proposed algorithm is *more precise*, while overcoming all shortcomings of the conventional methods.

Based on the proposed algorithm for generating correlated Rayleigh fading envelopes, the performance of our AST/DSTBC techniques proposed for systems utilizing DSTBCs in spatially correlated, flat Rayleigh fading channels is analyzed. Finally, the book examines the effect of imperfect carrier phase/frequency recovery at the receiver on the bit error performance of our AST/DSTBC schemes proposed for channels utilizing DSTBCs. The tolerance of differential detection associated with the proposed ASTs to phase/frequency errors is then analyzed. These analyses show that our ASTs not only work well in independent, flat Rayleigh fading channels as well as in the case of perfect carrier recovery, but also are very robust in correlated, flat Rayleigh fading channels as well as in the case of imperfect carrier recovery.

The book is concluded with recommendations on the issues examined here and with a number of future research directions.

LE CHUNG TRAN [1]

TADEUSZ A. WYSOCKI

ALFRED MERTINS

JENNIFER SEBERRY

---

[1] Le Chung Tran is also a lecturer at the Faculty of Electrical and Electronics, Hanoi University of Communications and Transport, Vietnam.

Acknowledgments

The authors would like to sincerely thank Prof. Joe Chicharo, Dean of Faculty of Informatics, University of Wollongong, Australia, Dr. Beata J. Wysocki, School of Electrical, Computer and Telecommunications Engineering, University of Wollongong, Australia, and Assistant Prof. Sarah A. Spence, Franklin Olin College of Engineering, USA for encouraging and helping us to write this book. We are also grateful to various colleagues who have enhanced our understandings of the subject, in particular to Y. Zhao. We would like to thank them for the enlightenment gained by our collaborations on papers, book chapters and projects, from which this book is the outcome. Last but not least, this book is dedicated to our families, who have enormously supported as well as mobilized us to write this book.

Chapter 1

OVERVIEW OF THE BOOK

1.1   Background

The first capacity theories, which extended Shannon's limit to the case of Multiple-Input Multiple-Output (MIMO) systems transmitting signals in multi-path channels, were derived by G. J. Foschini [Foschini, 1996] in 1996 and, concurrently, by Greg Raleigh and V. K. Jones [Griffith, 2004], [Group, 2003], [Jones et al., 2003]. These theories proved mathematically that the capacity and spectral efficiency of a MIMO system can be increased virtually indefinitely, without using extra frequency spectrum, by increasing the number of transmitter and receiver antennas. Since then, numerous researches and experimental measurements solidifying these theories have been derived in the literature. Digital communication using MIMO systems now becomes one of the most significant technical breakthroughs in modern communication. MIMO transmission is an outstanding technique with a chance to resolve the bottleneck of traffic capacity in the future wireless networks.

The main feature of MIMO systems is *space-time* processing. Space-Time Codes (STCs) are the codes designed for the use in MIMO systems. In STCs, signals are coded in both temporal and spatial domains. Among a variety of STCs, *orthogonal* Space-Time Block Codes (STBCs), which possess a much simpler decoding method over other STCs, are of particular interest in this book. Furthermore, *Complex Orthogonal* STBCs (CO STBCs) are *mainly* considered in this book since they can be used for PSK/QAM modulation schemes, and therefore, are more practical than real STBCs.

Particularly, this book discovers, proposes or examines the following issues:

- New constructions to improve the performance of CO STBCs;

- Multi-Modulation Schemes (MMSs) to increase the information transmission rates of CO STBCs;

- Methods to evaluate the optimal inter-symbol power allocation in multi-modulated CO STBCs;

- An improved diversity Antenna Selection Technique (AST) for channels using STBCs with *coherent detection*;

- Diversity ASTs for channels using differential STBCs (DSTBCs) with *differential detection* which are referred to as the AST/DSTBC schemes;

- A more general algorithm to generate correlated Rayleigh fading envelopes in wireless channels, including correlated MIMO channels;

- The robustness of the proposed AST/DSTBC schemes in correlated Rayleigh fading channels as well as in channels with imperfect carrier recovery.

The book deals with the above issues due to the fact that these issues are still open problems or even unknown in the literature so far and that, once they are resolved, they can provide significant improvements in MIMO systems using STBCs or DSTBCs in various aspects, such as information transmission rates, error performance, convenience in practical implementation.

1.2   Structure of the Book

The book contains eight main chapters which are outlined as follows:

- **Chapter 1**: In this chapter, the outline of the book is provided. The published and submitted papers, which are based on the book and merit the book, are mentioned here. Main contributions of the book are then discussed.

- **Chapter 2**: The background knowledge related to Multi-Input Multi-Output (MIMO) systems, Space-Time Block Codes (STBCs), and some conventional transmission diversity techniques is mentioned. This background knowledge is essential for the readers to understand the state of the art of the issues related to this book.

- **Chapter 3**: Three new, order-8, *maximum rate*, Complex Orthogonal Space-Time Block Codes (CO STBCs) with fewer zero entries, compared to the conventional CO STBCs of order 8, or even with no zero entries are proposed from the Clifford algebra. These codes require a lower peak-to-mean power ratio per transmitter antenna to achieve the same bit error rate characteristics as the conventional CO STBCs of order 8 with numerous zero entries. Equivalently, for the same peak power per transmitter antenna, the proposed CO STBCs achieve better bit error properties in comparison with the conventional CO STBCs. The derivation of our new CO STBCs can be presented by the milestones in the historical diagram of STBCs (see Fig. 1.1).

- **Chapter 4**: In this chapter, multi-modulation schemes (MMSs) are proposed to increase the data transmission rate of the two new CO STBCs proposed in Chapter 3 for eight transmitter antennas corresponding to the Amicable Orthogonal Designs (AODs) (8;1,1,2,2;1,1,2,2) and (8;1,1,1,4;1,1,1,4), respectively. In addition, the method to evaluate the optimal inter-symbol power allocation in the proposed codes in single modulation as well as in MMSs for both Additive White Gaussian Noise (AWGN) and flat Rayleigh fading channels is proposed. It turns out that, in some modulation schemes, equal power transmission per symbol time slot is not only optimal from the technical point of view, but also optimal in terms of achieving the best symbol error probability. The MMSs increasing the data transmission rate of CO STBCs and the optimal power allocation for multi-modulated CO STBCs mentioned in this chapter can be generalized for CO STBCs of other orders without any difficulty.

- **Chapter 5**: Constructions of square, *maximum rate* CO STBCs are well known, however, codes constructed via the known methods include numerous zeros, which impede their practical implementation. By modifying the Williamson and the Wallis-Whiteman arrays to apply to complex matrices, we propose two methods to construct the *square*, order-$4n$ CO STBCs from *square*, order-n CO STBCs satisfying certain properties. Applying the proposed methods, we construct *square*, *maximum rate*, order-8 CO STBCs with no zeros, such that the transmitted symbols equally disperse through transmitter antennas. Those codes, referred to as the *improved square CO STBCs*, have the advantages that the power is equally transmitted via each transmitter antenna during every symbol time slot and that a lower peak power per transmitter antenna is required to achieve the same bit error rates

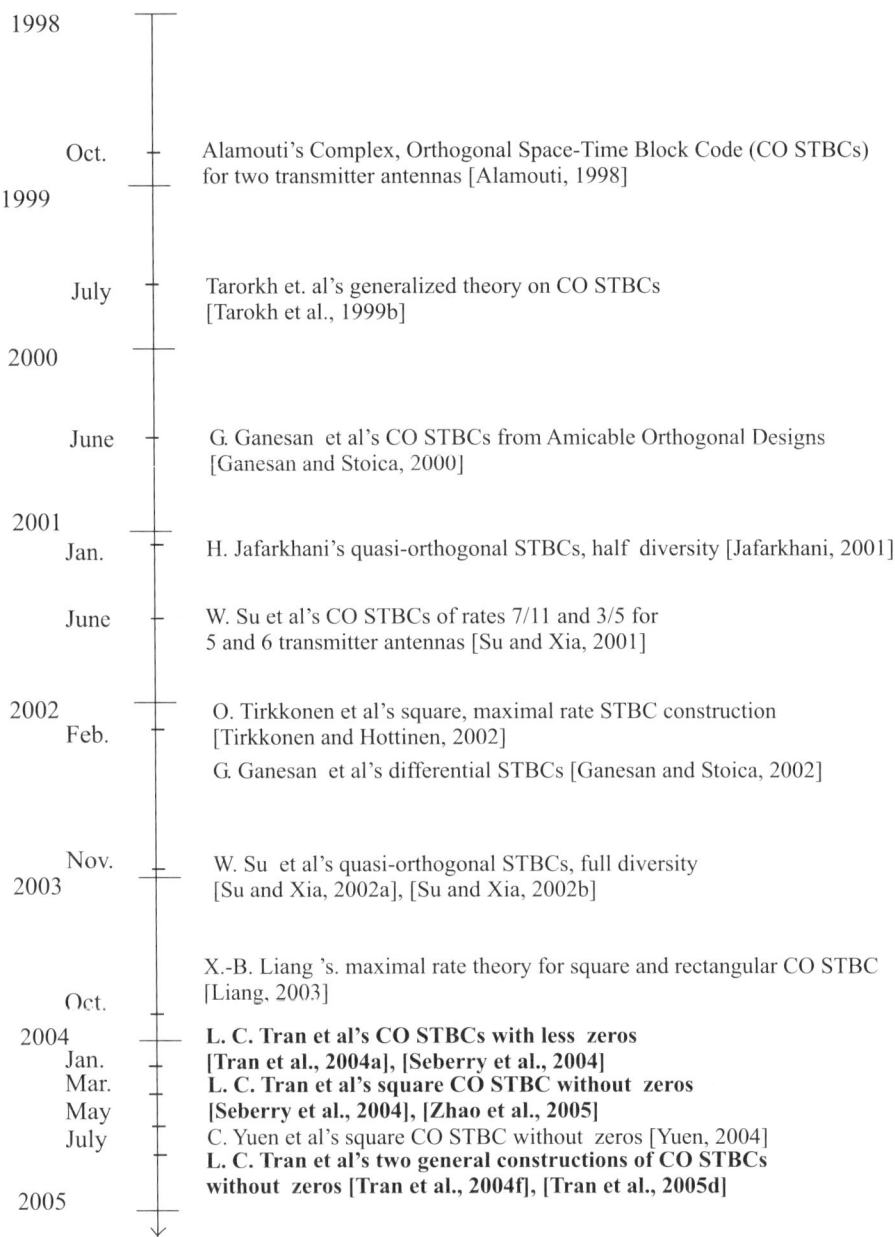

Figure 1.1.   History and main milestones of STBCs.

as in the conventional CO STBCs with zeros. The discovery of these new constructions is presented by another milestone in Fig. 1.1.

- **Chapter 6**: The combination of STBCs and a closed loop transmission diversity technique in both scenarios of *coherent* and *non-coherent* detection (*differential detection*) is considered in this chapter. This combination allows us to improve the performance of wireless communications channels.

  Firstly, we propose a simple closed loop transmission diversity technique to improve further the performance of channels using STBCs (with *coherent detection*), through proposing a structure of feedback information in order to reduce the time required for processing the feedback information at the transmitter. Calculations and simulations show that the proposed technique performs very well, especially, when it is combined with the Alamouti code.

  As opposed to the diversity antenna selection techniques (ASTs) for channels using STBCs with *coherent detection*, ASTs for channels using Differential Space-Time Block Codes (DSTBCs) with *differential detection* have *not* been well examined in the literature yet. The chapter thus proposes two ASTs for wireless channels utilizing DSTBCs, which remarkably improve performance of those channels.

- **Chapter 7**: In this chapter, a very general, straightforward algorithm for generation of an *arbitrary* number of Rayleigh envelopes with *any desired* (*equal* or *unequal*) power in wireless channels either *with* or *without* Doppler frequency shifts is proposed. The proposed algorithm can be applied to the case of *spatial correlation*, such as with antenna arrays in Multiple Input Multiple Output (MIMO) systems, or *spectral correlation* between the random processes like in Orthogonal Frequency Division Multiplexing (OFDM) systems. It can also be used for generating correlated Rayleigh fading envelopes in either *discrete-time instants* or a *real-time scenario*. Besides being *more generalized*, our proposed algorithm is *more precise*, while overcoming all shortcomings of the conventional methods.

  Based on the proposed algorithm, this chapter then analyzes the performance of our ASTs proposed for systems utilizing DSTBCs in spatially correlated, flat Rayleigh fading channels. The analysis shows that our ASTs not only work well in uncorrelated Rayleigh fading channels, but also are very robust in correlated channels.

This chapter also examines the effect of imperfect carrier phase/frequency recovery at the receiver on the bit error performance of our diversity ASTs proposed for channels utilizing DSTBCs. The tolerance of differential detection associated with the proposed ASTs to phase/frequency errors is then analyzed.

- **Chapter 8**: The conclusion of the book, recommendations, and discussion on the future works based on this book are provided here.

1.3   Contributions of the Book

The following contributions have resulted from this work:

1 Derivation of the most updated reviews on the capacity of MIMO systems, on the maximum rates of STBCs, on various constructions of square as well as non-square STBCs, and on the capacity of channels using STBCs in comparison with the real capacity of MIMO systems (Chapter 2).

2 Discovery of the three new, order-8 CO STBC with fewer zero entries and even with no zeros, which provide better properties in numerous aspects, compared to the conventional CO STBC of the same order (Chapter 3). Fig. 1.1 shows the milestones of this discovery in the historical diagram of STBCs. It is important to emphasize that we only consider *orthogonal SBTCs* in this book.

3 Proposition of Multi-Modulation Schemes (MMSs) to increase the data rates of the proposed CO STBCs (Section 4.3, Chapter 4).

4 Derivation of the method to examine the optimal power allocation between the symbols (or inter-symbol power allocation) in the proposed MMSs to achieve the improved error performance (Section 4.4, Chapter 4).

5 Observation that, in some MMSs, equal power transmission per symbol time slot is not only optimal from the technical point of view, but also optimal in terms of achieving the best symbol error probability (Section 4.4, Chapter 4).

6 Discovery of the two methods to construct the square CO STBCs having fewer zero entries, compared to the conventional CO STBCs, or even having no zero entries, which are referred to as the improved, square CO STBCs (Chapter 5). Fig. 1.1 shows the milestone of this discovery in the historical diagram of STBCs.

7 Derivation of the improved Antenna Selection Technique (AST) for channels using STBCs with coherent detection. The improved AST shortens the time required to process feedback information, which is used for selecting transmitter antennas, compared to the conventional AST (Section 6.2, Chapter 6).

8 Proposition of two main ASTs, namely, *general* $(M, N; K)$ AST/DSTBC scheme and *restricted* $(M, N; K)$ AST/DSTBC scheme, utilized for wireless channels using Differential Space-Time Block Codes (DSTBCs). These AST/DSTBC schemes improve significantly the bit error performance of channels using DSTBCs (Section 6.3, Chapter 6).

9 Derivation of a more generalized algorithm to generate correlated Rayleigh fading envelopes in the case of *either* spectral correlation as well as temporal delay, *or* spatial correlation (Section 7.2, Chapter 7).

10 Examination and affirmation of the fact that our ASTs proposed for channels using DSTBCs not only work very well in uncorrelated Rayleigh fading channels, but also are very robust in spatially, temporally or spectrally correlated ones (Section 7.3, Chapter 7).

11 Examination and affirmation of the robustness of the proposed ATSs in the imperfect carrier phase/frequency recovery scenario (Section 7.4, Chapter 7).

12 Statement of the open problems which are the potential researches for the topics related to this book (Chapter 8).

Chapter 2

MULTIPLE-INPUT MULTIPLE-OUTPUT SYSTEMS WITH SPACE-TIME CODES

2.1   Introduction

Digital communication using Multiple-Input Multiple-Output (MIMO) systems is one of the most significant technical breakthroughs in modern communication. MIMO systems are simply defined as the systems containing multiple transmitter antennas and multiple receiver antennas. Communication theories show that MIMO systems can provide a potentially very high capacity that, in many cases, grows approximately linear with the number of antennas. Recently, MIMO systems have already been implemented in wireless communication systems, especially in wireless LANs (Local Area Networks) [Griffith, 2004], [Group, 2003], [Jones et al., 2003]. Different structures of MIMO systems have also been proposed by industrial organizations in the Third Generation Partnership Project (3GPP) standardizations, including the structures proposed in [Electronics, 2004], [Ericsson, 2004], [Nokia, 2004], [Samsung and SNU, 2004]. The core idea under the MIMO systems is the ability to turn multi-path propagation, which is typically an obstacle in conventional wireless communication, into a benefit for users.

The main feature of MIMO systems is *space-time* processing. Space-Time Codes (STCs) are the codes designed for the use in MIMO systems. In STCs, signals are coded in both temporal and spatial domains. Among different types of STCs, *orthogonal* Space-Time Block Codes (STBCs) possess a number of advantages over other types of STCs (as mentioned in details later in this chapter) and are considered in this book.

In addition, the combination of STBCs and closed loop transmission diversity techniques using feedback loops has been investigated in the literature. When

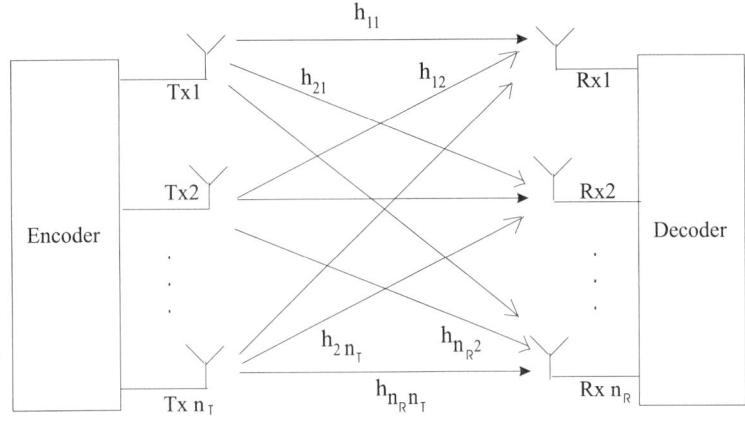

Figure 2.1. The diagram of MIMO systems.

applied, this combination improves significantly the performance of wireless systems. Several such transmission diversity techniques have been derived in the literature, such as space-time coded systems with beamforming [Nokia, 2002], antenna weighting [Electronics, 2002] or transmitter diversity antenna selection [Katz et al., 2001].

In Sections 2.2, 2.3 and 2.4 of this chapter, three main topics are respectively mentioned including:

■ MIMO systems from capacity perspectives;

■ Space-Time Block Codes;

■ Typical transmission diversity techniques

These topics are very important for readers to have the basic knowledge related to the issues mentioned in this book. Conclusions and research problems addressed in this book are mentioned in Section 2.5.

2.2   Multiple-Input Multiple-Output Wireless Communications

2.2.1   MIMO System Model

We consider a single user MIMO system comprising n_T transmitter antennas (n_T Tx antennas) and n_R receiver antennas (n_R Rx antennas). In particular, a complex baseband system described in discrete-time domain is of interest throughout the book. The block diagram of the MIMO system is presented in Fig. 2.1. During *each Symbol Time Slot* (STS), the transmitted signals are presented as an $n_T \times 1$ column vector $\mathbf{x}$, whose entry x_i, for $i = 1, \ldots, n_T$, is the transmitted signal at the ith Tx antenna during the considered STS.

We consider here an additive Gaussian channel (with or without Rayleigh fading) for which the optimal distribution of the transmitted signals in $\mathbf{x}$ is also Gaussian, i.e., the transmitted signals x_i, for $i = 1, \ldots, n_T$, are zero-mean, identically independently distributed (i.i.d.) complex random variables. The covariance matrix of $\mathbf{x}$ is

$$\mathbf{R}_{XX} = E\{\mathbf{x}\mathbf{x}^H\}$$

where $E\{.\}$ denotes the expectation, and $(.)^H$ denotes the Hermitian transposition operation. The total power of transmitted signals (during each STS) is constrained to P, regardless of the number of transmitter antennas n_T. It implies that

$$P = tr(\mathbf{R}_{XX})$$

where $tr(.)$ denotes the trace operation of the argument matrix.

In all following sections, we assume that channel coefficients (or transmission coefficients) are *perfectly known at the receiver*, but they *may or may not be known* at the transmitter. The scenario where channel coefficients are *unknown at both transmitter and receiver* is mentioned in [Marzetta and Hochwald, 1999]. Readers may refer to [Marzetta and Hochwald, 1999] for more details.

In the case where channel coefficients are unknown at the transmitter (but known at the receiver), we assume that the transmitted power at each Tx antenna is the same and equal to

$$P_{tj} = \frac{P}{n_T}$$

for $j = 1, \ldots, n_T$. In the case where the channel coefficients are known at the transmitter, the transmitted power is unequally assigned to the Tx antennas following the *water-filling* rule (see Appendix 1.1 in [Vucetic and Yuan, 2003]). We will mention this case in more details later in this chapter.

The channel is presented by an $n_R \times n_T$ complex matrix $\mathbf{H}$, whose elements h_{ij} are the channel coefficients between the jth Tx antenna ($j = 1, \ldots, n_T$) and the ith Rx antenna ($i = 1, \ldots, n_R$). Channel coefficients h_{ij} are assumed to be zero-mean, i.i.d. complex Gaussian random variables with a distribution $\mathcal{CN}(0, 1)$.

Noise at the receiver is presented by an $n_R \times 1$ column vector $\mathbf{n}$ whose elements are zero-mean, i.i.d. complex Gaussian random variables with identical variances (power) σ^2.

If we denote $\mathbf{r}$ to be the column vector of signals received at Rx antennas during each STS, then the transmission model is presented as

$$\mathbf{r} = \mathbf{Hx} + \mathbf{n}$$

If we assume that the average total power P_r received by each Rx antenna (regardless of noises) is equal to the average total transmitted power P from n_T Tx antennas, the Signal-to-Noise Ratio (SNR) at each Rx antenna is then

$$\rho = \frac{P_r}{\sigma^2} = \frac{P}{\sigma^2}$$

To guarantee the assumption that $P_r = P$, *for a channel with fixed channel coefficients* and with the *equal* transmitted power per Tx antenna P/n_T (i.e., in the case where channel coefficients are *known at the receiver*, but *unknown at the transmitter*), we must have the following constraint:

$$\sum_{j=1}^{n_T} |h_{ij}|^2 = n_T \tag{2.1}$$

for $i = 1, \ldots, n_R$. For a channel with *random* channel coefficients and with *equal* transmitted power per Tx antenna, the formula (2.1) is calculated with the expected value.

The *system capacity* $C(bits/s)$ is defined as the maximum possible transmission rate such that the error probability is arbitrarily small. In this book, we also consider the normalized capacity $C/W(bits/s/Hz)$, which is the system capacity C normalized to the channel bandwidth W.

2.2.2   Capacity of Additive White Gaussian Noise Channels with Fixed Channel Coefficients

In this section, at first, we derive the most general formula to calculate the channel capacity for both cases where channel coefficients are known as well as unknown at the transmitter. Based on this general formula, we will then derive the formulas for channel capacity in some particular cases.

The most general formula for calculating channel capacity in the case where channel coefficients are *either* known *or* unknown at the transmitter is the Shannon capacity formula (see Eq. (1.19) in [Vucetic and Yuan, 2003]):

$$C = W \sum_{i=1}^{r} \log_2 \left(1 + \frac{P_{ri}}{\sigma^2} \right) \tag{2.2}$$

where W is the bandwidth of each sub-channel, r is the rank of the channel coefficient matrix $\mathbf{H}$ (r is equal to the number of non-zero eigenvalues of $\mathbf{H}^H \mathbf{H}$), P_{ri} is the received power at each Rx antenna from the ith sub-channel, for $i = 1, \ldots, r$, during the considered symbol time slot. The term *"sub-channel"* is defined here as that mentioned in Section 1.3 of [Vucetic and Yuan, 2003]. Readers may refer to that section for more details. The rank r is at most equal to $m = \min(n_T, n_R)$.

2.2.2.1 Unknown Channel Coefficients at the Transmitter

In this case, as mentioned earlier, the transmitted power per Tx antenna is assumed to be identical and equal to $P_{tj} = P/n_T$. Let $\mathbf{Q}$ be the Wishart matrix defined as

$$\mathbf{Q} = \begin{cases} \mathbf{H}\mathbf{H}^H & \text{if } n_R < n_T \\ \mathbf{H}^H \mathbf{H} & \text{if } n_R \geq n_T \end{cases}$$

From Eq. (2.2), it has been proved (see Eq. (1.30) in [Vucetic and Yuan, 2003]) that the channel capacity for such a scenario is

$$C = W \, \log_2 \left[\det \left(\mathbf{I}_r + \frac{P}{n_T \sigma^2} \mathbf{Q} \right) \right] = W \, \log_2 \left[\det \left(\mathbf{I}_r + \frac{\rho}{n_T} \mathbf{Q} \right) \right] \quad (2.3)$$

where $\det(.)$ denotes the determinant of the argument matrix.

We consider some particular cases as follows:

- Single antenna channel: In this case, we have $r = n_T = n_R = 1$ and $\mathbf{Q} = h = 1$ (see Eq. (2.1)). From (2.3), the channel capacity is calculated as

$$C = W \, \log_2 \left[\det \left(1 + \frac{P}{\sigma^2} \right) \right] \quad (2.4)$$

At SNR $\rho = \frac{P}{\sigma^2} = 20dB$, for instance, the normalized capacity of the single antenna channel is $C/W = 6.658$ bits/s/Hz.

- Receive diversity: In this case, $n_T = 1$, $n_R \geq 2$ and $\mathbf{H} = (h_1 \ldots h_{n_R})^T$, where $(.)^T$ denotes the transposition operation. From (2.3), the channel capacity is calculated as

$$C = W \, \log_2 \left(1 + \frac{P}{\sigma^2} \sum_{i=1}^{n_R} |h_i|^2 \right)$$

Assuming that $|h_i|^2 = 1$, for $i = 1, \ldots, n_R$, then we have

$$C = W \, \log_2 \left(1 + \frac{P n_R}{\sigma^2}\right) \qquad (2.5)$$

For $n_R = 2$ and SNR $\rho = 20dB$, we have $C/W = 7.6511$ bits/s/Hz. We can see that the normalized capacity in this case is larger than that in the case of channels with single Tx and Rx antennas.

■ Transmit diversity: In this case, $n_T \geq 2$, $n_R = 1$ and $\mathbf{H} = (h_1 \ldots h_{n_T})$. From (2.3), the channel capacity is calculated as

$$C = W \, \log_2 \left(1 + \frac{P}{n_T \sigma^2} \sum_{i=1}^{n_T} |h_i|^2\right)$$

Assuming that $|h_i|^2 = 1$ for $i = 1, \ldots, n_T$, then we have

$$C = W \, \log_2 \left(1 + \frac{P}{\sigma^2}\right) \qquad (2.6)$$

From (2.6), we see that the capacity of the channel where *channel coefficients are fixed* and *unknown at the transmitter* is the same as that of the single antenna channel (see Eq. (2.4)), regardless of the number n_T of Tx antennas. Hence, for $n_T = 2$, $n_R = 1$ and SNR $\rho = 20dB$, we have $C/W = 6.658$ bits/s/Hz.

2.2.2.2   Known Channel Coefficients at the Transmitter

The channel capacity can be increased if channel coefficients are known at the transmitter. In this case, the transmitted power is assigned unequally to the Tx antennas, according to the "water-filling" rule, i.e., a larger power is assigned to a better sub-channel and visa versa (see Appendix 1.1 in [Vucetic and Yuan, 2003]). The power assigned to the ith sub-channel is

$$P_{ti} = \left(\mu - \frac{\sigma^2}{\lambda_i}\right)^+, \qquad i = 1, \ldots, r$$

where $(a)^+ = \max(a, 0)$, λ_i's are the non-zero eigenvalues of the matrix $\mathbf{H}^H \mathbf{H}$ (also $\mathbf{H}\mathbf{H}^H$) and μ is determined to satisfy the power constraint

$$\sum_{i=1}^{r} P_{ti} = P \qquad (2.7)$$

For the ith sub-channel, the received power P_{ri} at the receiver antenna is calculated as (see Eq. (1.20) in [Vucetic and Yuan, 2003]):

$$P_{ri} = \lambda_i P_{ti} = \left(\lambda_i \mu - \sigma^2\right)^+$$

Then, the channel capacity is derived from (2.2) as given below (see Eq. (1.35) in [Vucetic and Yuan, 2003]):

$$C = W \sum_{i=1}^{r} \log_2 \left[1 + \frac{(\lambda_i \mu - \sigma^2)^+}{\sigma^2} \right]$$

We consider the channel with $n_T \geq 2$ and $n_R = 1$ again. We have $r = \min(n_T, n_R) = 1$ and $\mathbf{H} = (h_1 \ldots h_{n_T})$. The power constraint (2.7) becomes

$$\mu - \frac{\sigma^2}{\lambda_1} = P$$

Equivalently, we have $\mu = P + \frac{\sigma^2}{\lambda_1}$, where the only eigenvalue λ_1 of the matrix $\mathbf{H}^H \mathbf{H}$ is $\lambda_1 = \sum_{i=1}^{n_T} |h_i|^2$. Therefore, we have

$$C = W \log_2 \left(1 + \frac{P \sum_{i=1}^{n_T} |h_i|^2}{\sigma^2} \right)$$

Assuming that $|h_i|^2 = 1$ for $i = 1, \ldots, n_T$, then we have

$$C = W \log_2 \left(1 + \frac{P n_T}{\sigma^2} \right)$$

For $n_T = 2$ and $SNR \ \rho = 20dB$, we have $C/W = 7.6511$ bits/s/Hz, which is larger than the channel capacity when the channel coefficients are *unknown* at the transmitter ($C/W = 6.658$ bits/s/Hz).

2.2.3 Capacity of Flat Rayleigh Fading Channels

In this scenario, channel coefficients are random, rather than being fixed as in Gaussian channels. We assume here that channel coefficients are zero-mean, i.i.d. complex Gaussian random variables with variances of 1/2 per dimension (real and imaginary). Hence, each channel coefficient has a Rayleigh distributed magnitude, uniformly distributed phase and the expected value of the squared magnitude equal to one, i.e., $E\{|h_{ij}|^2\} = 1$. We would like to stress that, in all following sections, channel coefficients are assumed to be *known at the receiver*, but *unknown at the transmitter*. Thus, the transmitted power per Tx antenna is assumed to be identical and equal to $P_{tj} = P/n_T$, for $j = 1, \ldots, n_T$.

We will consider the following scenarios, which have been widely mentioned in the literature, such as in [Telatar, 1999], [Vucetic and Yuan, 2003]:

■ The channel coefficient matrix $\mathbf{H}$ is random and its entries change randomly during every symbol time slot (STS). This scenario is referred to as the *fast, flat Rayleigh fading channel*.

- **H** is random and its entries change randomly after each block containing a fixed number of STSs. This scenario is referred to as the *block* flat Rayleigh fading channel.

- **H** is random but is selected at the beginning of transmission and its entries keep constant during the whole transmission. This scenario is referred to as the *slow* or *quasi-static*, flat Rayleigh fading channel.

We will consider the first two scenarios simultaneously in the following section.

2.2.3.1   Capacity of MIMO Systems in Fast and Block Rayleigh Fading Channels

It has been derived in the literature that the capacity of MIMO systems in fast and block Rayleigh fading channels is calculated as (see Eq. (1.56) in [Vucetic and Yuan, 2003] or Theorem 1 in [Telatar, 1999])

$$C = E\left\{ W \, \log_2\left[\det\left(\mathbf{I}_r + \frac{P}{n_T\sigma^2}\mathbf{Q} \right) \right] \right\} \tag{2.8}$$

where r is the rank of the matrix $\mathbf{H}$ and the matrix $\mathbf{Q}$ is the Wishart matrix defined as

$$\mathbf{Q} = \begin{cases} \mathbf{H}\mathbf{H}^H & \text{if } n_R < n_T \\ \mathbf{H}^H\mathbf{H} & \text{if } n_R \geq n_T \end{cases} \tag{2.9}$$

Eq. (2.8) can be evaluated with the aid of Laguerre polynomials (see Section 1.6.1 in [Vucetic and Yuan, 2003] or Theorem 2 in [Telatar, 1999])

$$\begin{aligned} C \; = \; & W \int_0^\infty \log_2\left(1 + \frac{P}{n_T\sigma^2}\lambda\right) \sum_{k=0}^{m-1} \frac{k!}{(k+n-m)!} \times \\ & \times \left[L_k^{n-m}(\lambda)\right]^2 \lambda^{n-m} e^{-\lambda} d\lambda \end{aligned} \tag{2.10}$$

where $n = \max(n_T, n_R)$, $m = \min(n_T, n_R)$ and

$$L_k^{n-m}(\lambda) = \frac{1}{k!}e^\lambda \lambda^{m-n} \frac{d^k}{d\lambda^k}(e^{-\lambda}\lambda^{n-m+k})$$

is the Laguerre polynomial of order k.

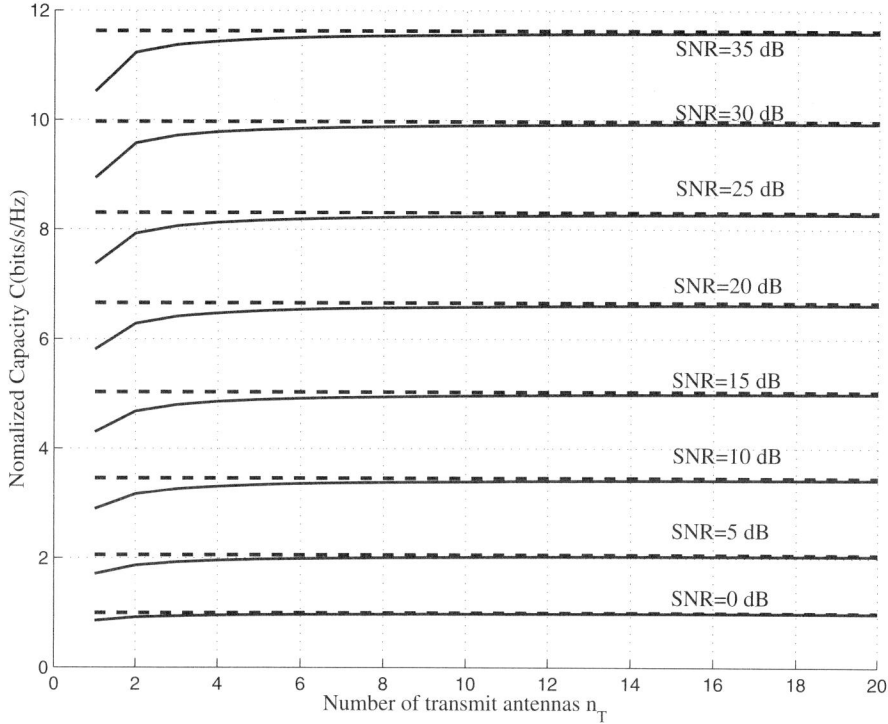

Figure 2.2. Capacity of MIMO systems with one Rx antenna in fast or block flat Rayleigh fading channels.

Furthermore, by increasing m and n, but keeping the ratio $\tau = \frac{n}{m} = const$, we have the following limit (see Eq. (1.61) in [Vucetic and Yuan, 2003] and Eq. (13) in [Telatar, 1999]):

$$\lim_{n \to \infty} \frac{C}{m} = \frac{W}{2\pi} \int_{\nu_1}^{\nu_2} \log_2 \left(1 + \frac{Pm}{n_T \sigma^2} \nu\right) \sqrt{\left(\frac{\nu_2}{\nu} - 1\right)\left(1 - \frac{\nu_1}{\nu}\right)}$$

where

$$\nu_1 = (\sqrt{\tau} - 1)^2$$

$$\nu_2 = (\sqrt{\tau} + 1)^2$$

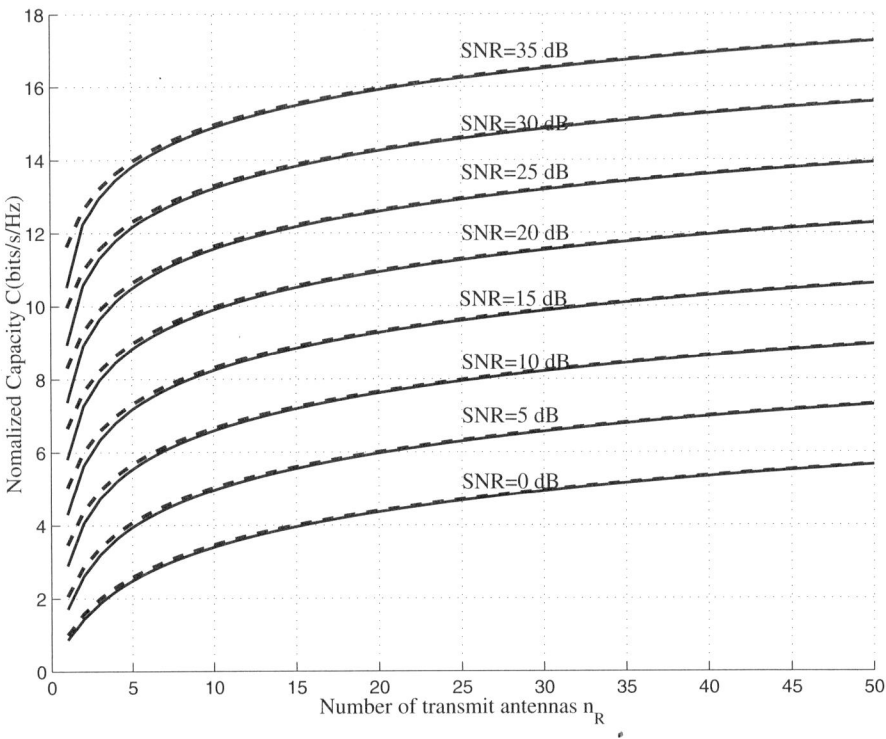

Figure 2.3. Capacity of MIMO systems with one Tx antenna in fast or block flat Rayleigh fading channels.

We consider now three scenarios:

■ Transmit diversity: In this case, we have $n_T \geq 2$ and $n_R = 1$. From (2.10), we have

$$C = W \frac{1}{(n_T - 1)!} \int_0^\infty \log_2 \left(1 + \frac{P}{n_T \sigma^2} \lambda\right) \lambda^{n_T - 1} e^{-\lambda} d\lambda \qquad (2.11)$$

When n_T increases, the capacity approaches the asymptotic value

$$\lim_{n_T \to \infty} C = W \log_2 \left(1 + \frac{P}{\sigma^2}\right) \qquad (2.12)$$

We realize that Eq. (2.12) is similar to Eq. (2.6). It means that, when the number of Tx antennas is large, the capacity of the transmit diversity system in *fast or block Rayleigh* fading channels approaches the capacity of the transmit diversity system in AWGN channels where channel coefficients are

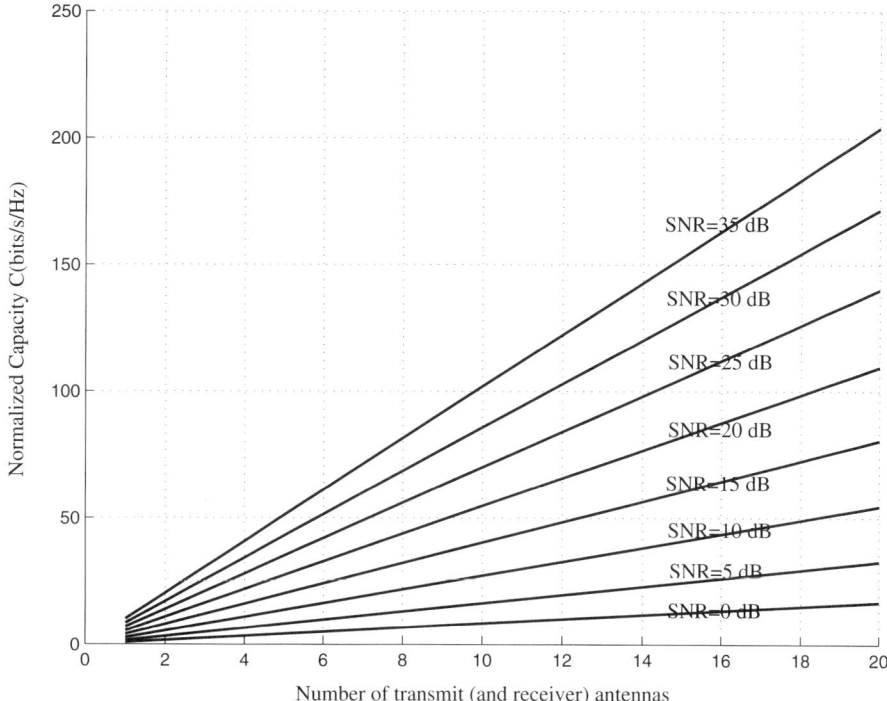

Figure 2.4. Capacity of MIMO systems with $n_T = n_R$ in fast or block flat Rayleigh fading channels.

unknown at the transmitter. The channel capacity of the transmit diversity systems is presented in Fig. 2.2. In this figure, the solid lines present the capacity calculated by (2.11), while the dashed lines present the asymptotic capacity calculated by (2.12). The dashed lines thus also present the capacity of AWGN channels calculated following (2.6).

- Receive diversity: In this case, we have $n_T = 1$ and $n_R \geq 2$. From (2.10), we have

$$C = W \frac{1}{(n_R - 1)!} \int_0^\infty \log_2 \left(1 + \frac{P}{\sigma^2}\lambda\right) \lambda^{n_R-1} e^{-\lambda} d\lambda \qquad (2.13)$$

When n_R increases, the capacity approaches the asymptotic value

$$\lim_{n_R \to \infty} C = W \log_2 \left(1 + \frac{Pn_R}{\sigma^2}\right) \qquad (2.14)$$

We realize that (2.14) is similar to (2.5). It means that, when the number of Rx antennas is large, the capacity of the receive diversity system in *fast or block Rayleigh* fading channels approaches the capacity of the receive diversity system in AWGN channels. Channel capacity of the receive diversity systems is presented in Fig. 2.3. The solid lines present the capacity calculated by (2.13), while the dashed lines present the asymptotic capacity calculated by (2.14). Similarly, the dashed lines also present the capacity of AWGN channels calculated following (2.5).

- Transmit and receive diversity: We assume further that $n_T = n_R$, and hence, $m = n = n_T = n_R$. Thus, from (2.10), channel capacity is calculated as

$$C = W \int_0^\infty \log_2 \left(1 + \frac{P\lambda}{n_R \sigma^2}\right) \sum_{k=0}^{n_R-1} [L_k^0(\lambda)]^2 e^{-\lambda} d\lambda \qquad (2.15)$$

where

$$L_k^0(\lambda) = \frac{1}{k!} e^\lambda \frac{d^k}{d\lambda^k} (e^{-\lambda} \lambda^k)$$

With the note that $m = n = n_T = n_R$, the empirical bound of the capacity in (2.15) has the following closed form (see Eq. (1.76) in [Vucetic and Yuan, 2003]):

$$\lim_{n \to \infty} \frac{C}{Wn} \geq \log_2 \left(\frac{P}{\sigma^2}\right) - 1$$

From this formula, it is clear that the capacity almost increases linearly with the number of Tx (and Rx) antennas. The capacity of the channel with $n_T = n_R$ is presented in Fig. 2.4.

2.2.3.2   Capacity of MIMO Systems in Slow Rayleigh Fading Channels

The results mentioned here were originally derived by Foschini and Gans [Foschini and Gans, 1998]. We consider a MIMO system where the channel coefficient matrix **H** is chosen randomly at the start of transmission and it stays constant during the whole transmission. The entries of **H** follow the Rayleigh distribution. Wireless Local Area Networks (LANs) with high data rates and low fade rates are examples of this scenario.

Again, we consider three cases as follows:

- Receive diversity: In this case, we have $n_T = 1$ and $n_R \geq 2$. It has been shown by Eq. (10) in [Foschini and Gans, 1998] or by Eq. (1.78) in [Vucetic and Yuan, 2003] that the channel capacity is calculated as

$$C = W \log_2 \left(1 + \frac{P}{\sigma^2} \chi_{2n_R}^2 \right)$$

where $\chi_{2n_R}^2$ is a chi-square random variable with $2n_R$ degrees of freedom.

- Transmit diversity: In this case, we have $n_T \geq 2$ and $n_R = 1$. It has been shown by Eq. (11) in [Foschini and Gans, 1998] or by Eq. (1.79) in [Vucetic and Yuan, 2003] that the channel capacity is calculated as

$$C = W \log_2 \left(1 + \frac{P}{n_T \sigma^2} \chi_{2n_T}^2 \right)$$

where $\chi_{2n_T}^2$ is a chi-square random variable with $2n_T$ degrees of freedom.

- Transmit and receive diversity: We further assume that $n = n_T = n_R$ and n is large, then it is shown by Eq. (20) in [Foschini and Gans, 1998] or by Eq. (1.82) in [Vucetic and Yuan, 2003] that the lower bound on the capacity is

$$\frac{C}{Wn} > \left(1 + \frac{\sigma^2}{P} \right) \log_2 \left(1 + \frac{P}{\sigma^2} \right) - \log_2 e + \varepsilon_n \qquad (2.16)$$

where ε_n is a Gaussian random variable with the mean and variance as given below:

$$E\{\varepsilon_n\} = \frac{1}{n} \log_2 \left(1 + \frac{P}{\sigma^2} \right)^{-1/2}$$

$$Var\{\varepsilon_n\} = \left(\frac{1}{n \ln 2} \right)^2 \left[\ln \left(1 + \frac{P}{\sigma^2} \right) - \frac{\frac{P}{\sigma^2}}{1 + \frac{P}{\sigma^2}} \right]$$

From (2.16), we realize that when the number of Tx (and Rx) antennas is large, the channel capacity is linearly proportional to the number of antennas. For SNR $\rho = \frac{P}{\sigma^2} = 20dB$, and $n = n_T = n_R = 8$, the normalized capacity is $C/W \approx 37$ bits/s/Hz.

2.3   Space-Time Block Codes

Communication requires a very high rate with high reliability these days. Two major difficulties to obtain reliable communication via high rate wireless communication systems are bandwidth limitation of communication channels and multipath fading. To surmount these difficulties, multiple antenna systems referred to as MIMO systems, which provide a transmit and/or receive diversity, can be used. As mentioned in Section 2.2 of this book or in [Foschini, 1996], [Foschini and Gans, 1998], [Marzetta and Hochwald, 1999], [Telatar, 1999], MIMO systems can provide a potentially huge capacity gain with the same requirements for power and bandwidth as the single antenna systems. In many cases, the capacity of channels is proved to increase linearly with the lower number among the number of transmitter antennas (Tx antennas) and that of receiver antennas (Rx antennas).

Based on those information theoretical results, various schemes for the transmission of signals via MIMO systems have been proposed, including Bell Lab Layered Space-Time (BLAST) [Foschini, 1996], Space-Time Trellis Codes (STTCs) [Tarokh et al., 1998], Space-Time Block Codes (STBCs) [Alamouti, 1998], [Tarokh et al., 1999a], [Tarokh et al., 1999b] and Unitary Space-Time Codes [Hochwald and Marzetta, 2000] among many others. All designs are targeted to the transmission of signals in MIMO systems to achieve diversity and data rates as high as possible, while bandwidth expansion (if any) must be kept as small as possible.

Particularly, Tarokh et al. [Tarokh et al., 1998] proposed a few Space-Time Trellis Codes (STTCs) for 2–4 Tx antennas, which perform well in slow fading environment and have almost no loss of capacity compared to the channel capacity. However, the complexity of decoders increases exponentially with the size of signal constellations. The authors in [Tarokh et al., 1998] also derived the criteria for designing a good Space-Time Code (STC), including rank criterion and determinant criterion.

Later, Alamouti [Alamouti, 1998] discovered a very simple transmitter diversity technique for two Tx antennas, which provides a full diversity order, has no loss of capacity (if the number of Rx antennas is equal to one), and possesses a simple and fast maximum likelihood (ML) decoding. Instead of being joined, the transmitted signals are decoded separately at decoders due to the orthogonality between the columns (and rows) of the code. The Alamouti code could be considered as the *original* Space-Time Block Code (STBC) - the Space Time Codes (STCs) constructed from orthogonal designs. The discovery of the Alamouti code is presented by a milestone in Fig. 1.1 of Chapter 1.

Although some other designs were proposed for STCs, such as non-orthogonal designs based on number theory [A.-Meraim et al., 2002], [Damen et al., 2002], orthogonal STBCs are currently receiving an intensive attention due to the following reasons:

1 They posses a fast and very simple maximum likelihood decoding [Tarokh et al., 1999a], [Tarokh et al., 1998] due to their orthogonality.

2 They provide a full diversity order for a certain number of Tx antennas. Consequently, these codes have good error probability characteristics [Tarokh et al., 1999b].

Motivated by the Alamouti code, Tarokh et al. [Tarokh et al., 1999a], [Tarokh et al., 1999b] proposed STBCs for various numbers of Tx antennas. Based on signal constellations, the authors classified STBCs into two classes, namely, STBCs for real signals and STBCs for complex signals. Real STBCs can be used in the case of Pulse Amplitude Modulation (PAM) while complex STBCs are used for Phase Shift Keying (PSK) or Quadrature Amplitude Modulation (QAM) constellations. Both of them may, or may not, include linear processing (LP) at transmitters. The term "linear processing" will be explained in more details later.

Real STBCs have been well examined. There is a systematic method to construct real STBCs with the maximum rate $R_{max} = 1$ for up to 8 Tx antennas based on Huwitz-Radon theory. The background on the Huwitz-Radon theory can be found in [Ganesan and Stoica, 2000], [Geramita and Seberry, 1979]. These codes also provide a maximum Signal-to-Noise Ratio (SNR) at receivers [Ganesan and Stoica, 2001].

Unlike real STBCs, complex STBCs have not been well known in the literature yet, while they are more practical. Complex STBCs have recently received much attention. For these reasons, in this book, we will mainly focus on complex orthogonal STBCs (CO STBCs).

In this section, we present the basic theories on STBCs which are mainly based on the contributions of Alamouti [Alamouti, 1998], Liang [Liang, 2003], and Tarokh et al. [Tarokh et al., 1999a], [Tarokh et al., 1999b], [Tarokh et al., 1999c], [Tarokh et al., 1998].

A STBC representing a relationship between original transmitted symbols s_i and their replicas artificially created by the transmitter for transmission over the channel with multiple Tx antennas is defined by a $p \times n_T$ matrix, where p is the number of symbol time slots (STSs) for transmission of one code block and n_T is the number of Tx antennas. Generally, the elements of the matrix are linear

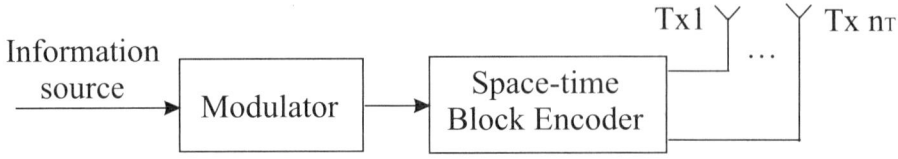

Figure 2.5. Space-time block encoding.

combinations of k input symbols s_i (i=1...k), which represent the information-bearing binary bits to be transmitted. Assuming that the signal constellation consisting of 2^b points is considered. Then, b binary bits are represented by a symbol s_i. Therefore, a block of $k \times b$ binary bits is entered into the encoder at a time and the encoding process is carried out in both space and time, hence, the code block at the output of the encoder is referred to as a Space-Time Block Codes (STBC). The space-time encoding process is presented by Fig. 2.5.

This block of the transmitted symbols is mathematically represented by a matrix $\mathbf{X}$ of size $p \times n_T$ as follows:

$$\begin{bmatrix} g_{11} & g_{12} & \cdots & g_{1n_T} \\ g_{21} & g_{22} & \cdots & g_{2n_T} \\ \cdots & \cdots & \cdots & \cdots \\ g_{p1} & g_{p2} & \cdots & g_{pn_T} \end{bmatrix}$$

where g_{jl}, for $j = 1, \ldots, p$ and $l = 1, \ldots, n_T$, represents a linear combination of the symbols s_i. The entries $g_{j1}, \ldots, g_{jn_T}$ are transmitted simultaneously from n_T Tx antennas at the jth time slot. Clearly, the length p of the STBC represents the delay for transmission (and for decoding as well). Since k symbols are transmitted during p STSs, the code rate of the STBC is defined by the ratio

$$R = k/p$$

The code rate is related to the spectral efficiency of the STBC as given below:

$$\eta = \frac{r_b}{B} = \frac{r_s b}{B} \qquad bits/s/Hz$$

where B denotes the bandwidth, r_b and r_s denote the bit rate and the symbol rate, respectively. The bandwidth B is calculated as

$$B = \frac{r_s}{R} = \frac{r_s p}{k} \qquad Hz \qquad (2.17)$$

Therefore, the spectral efficiency of the STBC is

$$\eta = \frac{r_b}{B} = bR = \frac{bk}{p} \qquad bits/s/Hz$$

Clearly, for the same signal constellation, a higher code rate results in a more efficient STBC. In other words, the code rate R represents the spectral efficiency of the STBC.

From Eq. (2.17), we also realize that if R is smaller than one, then there exists a bandwidth expansion to transmit STBCs, compared to, for instance, the transmission without using STBCs where each symbol is transmitted during each STS via one Tx antenna. It will be analyzed in more details later that the Alamouti STBC [Alamouti, 1998] achieves the full rate, i.e. $R = 1$, and consequently, does not expand transmission bandwidth.

It should be emphasized that, in the literature, STBCs usually refer to *orthogonal STBCs*. Other classes of STCs include *quasi-orthogonal* STBCs [Jafarkhani, 2001], [Su and Xia, 2002a], [Su and Xia, 2002b], Space-Time Trellis Codes (STTCs) [Tarokh et al., 1998], Layered Space-Time Codes (LSTs) [Foschini, 1996], [Vucetic and Yuan, 2003], or Linear Space-Time Codes [Hassibi and Hochwald, 2001], [Hassibi and Hochwald, 2002]. Orthogonal STBCs are of great interest as decoding processes of those codes only involve the linear processing at the receiver thanks to the orthogonality between their columns, and consequently, it is very simple to decode them. Because of this historical reason, the term "STBCs" should be understood as *orthogonal* STBCs throughout this book.

The main milestones in examining STBCs so far are presented in Fig. 1.1 in Chapter 1, which is explained in more details by the following sections.

2.3.1   Real Orthogonal Designs

DEFINITION 2.3.1 *A generalized, real orthogonal design (also called generalized, real, orthogonal STBC) $\mathcal{G}$ is defined as a $p \times n$ matrix whose nonzero entries are the indeterminates x_1, x_2, ..., x_k over the real number field $\mathbb{R}$ or their negatives $-x_1, -x_2, \ldots, -x_k$ such that*

$$\mathcal{G}^T \mathcal{G} = \mathcal{D} \qquad (2.18)$$

where $\mathcal{G}^T$ denotes the transpose of the matrix $\mathcal{G}$, while $\mathcal{D}$ is a diagonal matrix of size $n \times n$ with diagonal entries $\mathcal{D}_{ii}$, for $i = 1, 2, \ldots, n$, of the form $(l_{i1}x_1^2 + l_{i2}x_2^2 + \cdots + l_{ik}x_k^2)$ and coefficients $l_{i1}, \ldots, l_{ik}$ are strictly positive, real numbers. The rate of $\mathcal{G}$ is $R = k/p$. The matrix $\mathcal{G}$ is said to be a $[p, n, k]$

real orthogonal design. If $p = n$, $\mathcal{G}$ is called square, real orthogonal design. If the coefficients $l_{i1}, \ldots, l_{ik}$ satisfy $l_{i1} = \cdots = l_{ik} = 1$, for $i = 1, 2, \ldots, n$, then $\mathcal{G}$ is called real, orthogonal STBC without Linear Processing (LP). Otherwise, $\mathcal{G}$ is called real, orthogonal STBC with LP.

The condition on the positive definiteness of the coefficients l_{ij}s is to guarantee that the STBC $\mathcal{G}$ provides a full diversity order.

Proof. We prove that $\mathcal{G}$ satisfies the rank criterion for designing STBCs, i.e. [Tarokh et al., 1998]:

$$\det \left[\mathcal{G}(\acute{x}_1 - x_1, \acute{x}_2 - x_2, \ldots, \acute{x}_k - x_k)^T \times \right.$$
$$\left. \times \mathcal{G}(\acute{x}_1 - x_1, \acute{x}_2 - x_2, \ldots, \acute{x}_k - x_k) \right] \neq 0$$

for any distinct pair of codewords $\acute{\mathbf{x}} \triangleq (\acute{x}_1, \ldots, \acute{x}_k)$ and $\mathbf{x} \triangleq (x_1, \ldots, x_k)$. Note that $\mathcal{G}(\acute{x}_1 - x_1, \acute{x}_2 - x_2, \ldots, \acute{x}_k - x_k)$ is the matrix $\mathcal{G}(x_1, x_2, \ldots, x_k)$ when we replace the set $(x_1, x_2, \ldots, x_k)$ by the set $(\acute{x}_1 - x_1, \acute{x}_2 - x_2, \ldots, \acute{x}_k - x_k)$. From (2.18), we have

$$\det \left(\mathcal{G}^T \mathcal{G} \right) = \prod_{i=1}^{n} \left[\sum_{j=1}^{k} l_{ij} x_j^2 \right]$$

Therefore

$$\det \left[\mathcal{G}(\acute{x}_1 - x_1, \acute{x}_2 - x_2, \ldots, \acute{x}_k - x_k)^T \times \right.$$
$$\left. \times \mathcal{G}(\acute{x}_1 - x_1, \acute{x}_2 - x_2, \ldots, \acute{x}_k - x_k) \right]$$
$$= \prod_{i=1}^{n} \left[\sum_{j=1}^{k} l_{ij} (\acute{x}_j - x_j)^2 \right]$$

Evidently, if l_{ij}s are positive definite, for any distinct pair of codewords $\acute{\mathbf{x}}$ and $\mathbf{x}$, we always have

$$\det \left[\mathcal{G}(\acute{x}_1 - x_1, \acute{x}_2 - x_2, \ldots, \acute{x}_k - x_k)^T \times \right.$$
$$\left. \times \mathcal{G}(\acute{x}_1 - x_1, \acute{x}_2 - x_2, \ldots, \acute{x}_k - x_k) \right] > 0$$

It means that, $\mathcal{G}$ satisfies the design criterion concerning to rank. Hence, the STBC $\mathcal{G}$ provides a full diversity order. ■

Real STBCs can be used for any Pulse Amplitude Modulation (PAM) or may even be used for Binary Phase Shift Keying (BPSK) and, generally, they are only used for those modulation techniques. In addition, real STBCs have been *well examined* in the literature.

It is more convenient to express $p = 2^{4c+d}.q$ where $0 \le c, 0 \le d < 4$, q is odd, and c, d, q are in the natural number field $\mathbb{N}$. Furthermore, let $A(R, n) = p_{min}$ be the minimum number p such that there exists a generalized orthogonal design of size $p \times n$ providing a code rate of at least $R=k/p$ (if there is no such design, then $A(R, n) = \infty$). It is proven in [Tarokh et al., 1999b] (pp.1461) that:

1 For any R, $A(R, n) < \infty$. It means that for any code rate R (and, therefore, including $R = 1$), generalized, real orthogonal designs always exist.

2 Full-rate, real STBCs exist for any number of n.

3 For any generalized orthogonal design of full rate ($R = 1$), one has $A(1, n) = p_{min} = \min(2^{4c+d})$, where the minimization is taken from the set

$$\{c, d \mid c, d \in \mathbb{N}, \ 0 \le c, \ 0 \le d < 4, \ \rho(p) \overset{\Delta}{=} 8c + 2^d \ge n\}$$

In the above equation, $\rho(p)$ is referred to as the Hurwitz-Radon number, which will be defined in more details later. The minimum length of any *full-rate*, real orthogonal STBCs, i.e., $A(1, n)$, can be determined via the condition on the Hurwitz-Radon number $\rho(p)$, which is stated by the following theorem (see Proposition 4 in [Liang, 2003]).

THEOREM 2.3.2 *For any number of n, a rate-1, $[p, n, p]$ real orthogonal STBC exists if and only if (iff) the Hurwitz-Radon number $\rho(p) \ge n$, i.e.:*

$$\rho(p) \overset{\Delta}{=} 8c + 2^d \ge n$$

The value of $A(1, n)$ presents the minimum length of the full-rate STBCs and also presents the minimum requirement on memory to achieve the full rate. Some values of $A(1, n)$ are presented in Table 2.1.

If we consider n as the number of Tx antennas and p as the delay of STBCs, then real orthogonal designs providing a full rate exist for any number n of Tx antennas. In addition, the optimal delay for real STBCs for 1, 2, 4 and 8 Tx antennas are 1, 2, 4 and 8 symbol time slots, respectively. In other words, the *square*, real STBCs only exist for 1, 2, 4 and 8 Tx antennas.

In general, for full-rate, real STBCs, $A(1, n)$ (or p_{min}) has the following form [Tirkkonen and Hottinen, 2002]:

$$p_{min} = A(1, n) = 16^{\lfloor (n-1)/8 \rfloor} 2^{\lceil log_2(1+(n-1) mod 8) \rceil}$$

Table 2.1. Some typical values of p_{min} (or $A(1, n)$) for the *full-rate, real* STBCs.

n	c	d	p_{min}	n	c	d	p_{min}
1	0	0	1	9	1	0	16
2	0	1	2	10	1	1	32
3	0	2	4	11	1	2	64
4	0	2	4	12	1	2	64
5	0	3	8	13	1	3	128
6	0	3	8	14	1	3	128
7	0	3	8	15	1	3	128
8	0	3	8	16	1	3	128

where $\lfloor a \rfloor$ is the smallest integer which is equal to or greater than a, and $\lceil a \rceil$ is the biggest integer which is equal to or smaller than a.

Some examples of full-rate, generalized orthogonal designs are given below:
For $n=1$, $p=1$:

$$\mathcal{G}_1 = (x_1) \tag{2.19}$$

For $n=2$, $p=2$:

$$\mathcal{G}_2 = \begin{bmatrix} x_1 & x_2 \\ -x_2 & x_1 \end{bmatrix} \tag{2.20}$$

For $n=3$, $p=4$:

$$\mathcal{G}_3^T = \begin{bmatrix} x_1 & -x_2 & -x_3 & -x_4 \\ x_2 & x_1 & x_4 & -x_3 \\ x_3 & -x_4 & x_1 & x_2 \end{bmatrix}$$

For $n=4$, $p=4$:

$$\mathcal{G}_4 = \begin{bmatrix} x_1 & x_2 & x_3 & x_4 \\ -x_2 & x_1 & -x_4 & x_3 \\ -x_3 & x_4 & x_1 & -x_2 \\ -x_4 & -x_3 & x_2 & x_1 \end{bmatrix} \tag{2.21}$$

For $n=5$, $p=8$:

$$
G_5^T = \begin{bmatrix}
x_1 & -x_2 & -x_3 & -x_4 & -x_5 & -x_6 & -x_7 & -x_8 \\
x_2 & x_1 & -x_4 & x_3 & -x_6 & x_5 & x_8 & -x_7 \\
x_3 & x_4 & x_1 & -x_2 & -x_7 & -x_8 & x_5 & x_6 \\
x_4 & -x_3 & x_2 & x_1 & -x_8 & x_7 & -x_6 & x_5 \\
x_5 & x_6 & x_7 & x_8 & x_1 & -x_2 & -x_3 & -x_4
\end{bmatrix}
$$

For $n=8$, $p=8$:

$$
G_8 = \begin{bmatrix}
x_1 & x_2 & x_3 & x_4 & x_5 & x_6 & x_7 & x_8 \\
-x_2 & x_1 & x_4 & -x_3 & x_6 & -x_5 & -x_8 & x_7 \\
-x_3 & -x_4 & x_1 & x_2 & x_7 & x_8 & -x_5 & -x_6 \\
-x_4 & x_3 & -x_2 & x_1 & x_8 & -x_7 & x_6 & -x_5 \\
-x_5 & -x_6 & -x_7 & -x_8 & x_1 & x_2 & x_3 & x_4 \\
-x_6 & x_5 & -x_8 & x_7 & -x_2 & x_1 & -x_4 & x_3 \\
-x_7 & x_8 & x_5 & -x_6 & -x_3 & x_4 & x_1 & -x_2 \\
-x_8 & -x_7 & x_6 & x_5 & -x_4 & -x_3 & x_2 & x_1
\end{bmatrix}
\tag{2.22}
$$

2.3.1.1 Maximum Rates of *Square, Real* Orthogonal Designs

As mentioned before, there always exist the real orthogonal designs $[p, n, k]$ with a full rate for any number of n. However, a question which could be raised is what the maximum number k of variables, say k_{max}, in the *square, real* orthogonal designs is. To answer this question, we need to define the Hurwitz-Radon number $\rho(n)$ which has been intensively mentioned in the literature, such as [Geramita and Seberry, 1979], [Liang, 2003], [Tarokh et al., 1999b].

DEFINITION 2.3.3 *If $n = 2^a(2b + 1)$ and $a = 4c + d$, where a, b, c and d are integers with $0 \leq d < 4$, then the Hurwitz-Radon number is defined as $\rho(n)=8c + 2^d$. $\rho(n)$ can be rewritten as follows:*

$$
\rho(n) = \rho(2^a(2b+1)) = \begin{cases}
2a + 1 & \text{if } a = 0 \ (mod) \ 4 \\
2a & \text{if } a = 1 \ (mod) \ 4 \\
2a & \text{if } a = 2 \ (mod) \ 4 \\
2a + 2 & \text{if } a = 3 \ (mod) \ 4
\end{cases}
\tag{2.23}
$$

where mod denotes the modulo operation.

Table 2.2. The maximum number of variables and the maximum rates of *square, real* STBCs.

n	k_{max}	R_{Rmax}	n	k_{max}	R_{Rmax}
1	1	1	9	1	1/9
2	2	1	10	2	1/5
3	1	1/3	11	1	1/11
4	4	1	12	4	1/3
5	1	1/5	13	1	1/13
6	2	1/3	14	2	1/7
7	1	1/7	15	1	1/15
8	8	1	16	9	9/16

The Hurwitz-Radon numbers have the following properties:

$$\rho(2^a(2b+1)) = \rho(2^a)$$
$$\rho(16n) = \rho(n) + 8$$
$$\rho(2^a) < \rho(2^{a+1})$$

It has been proved that the maximum number k_{max} of variables in a square, real orthogonal design is equal to the Hurwitz-Radon number $\rho(n)$, i.e., [Adams et al., 1965], [Geramita and Seberry, 1979], [Liang, 2003]:

$$k_{max} = \rho(n) = \rho(2^a) = \rho(2^a(2b+1))$$
$$= 8c + 2^d \tag{2.24}$$

From Eq. (2.24), we have the following corollary:

COROLLARY 2.3.4 *The maximum rate of square, real orthogonal designs for any number of Tx antennas* $n = 2^{4c+d}(2b+1)$, *denoted by* R_{Rmax}, *is calculated as*

$$R_{Rmax} = \frac{k_{max}}{n} = \frac{\rho(n)}{n} = \frac{8c + 2^d}{n} \tag{2.25}$$

In (2.25), the subscript "R" implies that the formula is applied to *real* designs. The maximum numbers of variables and the maximum rates of *square, real* STBCs for some typical values of n are given in Table 2.2. From Eq. (2.25) and Table 2.2, some following important notes are derived:

1 If n is odd, the maximum code rate of square, real orthogonal designs is $R_{Rmax} = 1/n$. Therefore, for $n > 3$, the code rate is very small, and hence,

there might be no point to examine *square, real* STBCs for the odd number of Tx antennas which is greater than 3.

2 *Square, real* STBCs with a full rate exist only for $n = 1, 2, 4$ and 8 Tx antennas. Note that the full-rate, real STBCs exist for any number n of Tx antennas, but for $n \neq 1, 2, 4$ or 8, those full-rate, real STBCs *must be non-square*.

2.3.1.2 Constructions of Maximum Rate, Square, Real STBCs

In this section, three methods for the construction of maximum rate, square, real STBCs are presented, including 1) the Adams-Lax-Phillips construction from octonions; 2) the Adams-Lax-Phillips construction from quaternions; and 3) the Geramita-Pullman construction. A good summary of these constructions can be found in Liang's paper [Liang, 2003].

1 *The Adams-Lax-Phillips construction from octonions [Adams et al., 1965], [Adams et al., 1966]:* We already have 4 square, real STBCs for $n = 2^a$ with $a = 0, 1, 2$ and 3, i.e., Eq. (2.19) for $a = 0$; Eq. (2.20) for $a = 1$; Eq. (2.21) for $a = 2$ and Eq. (2.22) for $a = 3$. Denote

$$\mathcal{G}_{2^a} = \mathcal{G}_{2^a}(x_1, \ldots, x_{\rho(n)}) \tag{2.26}$$

which has $\rho(n) = \rho(2^a)$ real variables and is of size $n \times n$.

From $\mathcal{G}_{2^a}$, one can construct a square, real STBC of order 2^{a+1} comprising $\rho(n) + 1$ real variables as follows:

$$
\begin{aligned}
\mathcal{G}_{2^{a+1}} &= \mathcal{G}_{2^{a+1}}(x_1, \ldots, x_{\rho(n)+1}) \\
&= \begin{bmatrix} x_{\rho(n)+1}\mathbf{I}_n & \mathcal{G}_{2^a} \\ \mathcal{G}_{2^a}^T & -x_{\rho(n)+1}\mathbf{I}_n \end{bmatrix}
\end{aligned} \tag{2.27}
$$

where $\mathbf{I}_n$ denotes the identity matrix of order n.

From $\mathcal{G}_{2^{a+1}}$, one can construct a square, real STBC of order $16n = 2^{a+4}$ comprising $\rho(n) + 8$ real variables as follows:

$$
\begin{aligned}
\mathcal{G}_{2^{a+4}} &= \mathcal{G}_{2^{a+4}}(x_1, \ldots, x_{\rho(n)+8}) \\
&= \mathcal{G}_{2^{a+1}} \bigotimes \mathbf{I}_8 + \mathbf{I}_{2n} \bigotimes \mathcal{G}_8(0, x_{\rho(n)+2}, \ldots, x_{\rho(n)+8})
\end{aligned} \tag{2.28}
$$

where $\mathcal{G}_8$ is defined by (2.22) and $\otimes$ denotes the Kronecker product.

From (2.26) – (2.28), the transition from order $n = 2^a$ to order $16n$ can be expressed as

$$\mathcal{G}_{2^{a+4}} = \begin{bmatrix} \mathbf{I}_n \otimes \mathcal{G}_8(y_1, y_2, \ldots, y_8) & \mathcal{G}_{2^a} \otimes \mathbf{I}_8 \\ \mathcal{G}_{2^a}^T \otimes \mathbf{I}_8 & \mathbf{I}_n \otimes \mathcal{G}_8(-y_1, y_2, \ldots, y_8) \end{bmatrix}$$

where $y_i = x_{\rho(n)+i}$ for $i = 1, \ldots, 8$.

In order to construct the square, real STBC of order $n = 2^a(2b + 1)$, from the square, real STBC of order $n = 2^a$, say $\mathcal{G}_{2^a}$, we just need to perform the Kronecker product between $\mathbf{I}_{(2b+1)}$ and $\mathcal{G}_{2^a}$.

The above constructions can be used to generate a maximum rate, square, real orthogonal design of order n with $\rho(n)$ real variables for any number of $n \in \mathbb{N}$ from the initial four orthogonal designs of orders $n = 1, 2, 4, 8$.

2 *The Adams-Lax-Phillips construction from quaternions [Adams et al., 1965]*: Another construction method using quaternions was introduced in [Adams et al., 1965]. This construction is another method for the transition from orthogonal designs of order $n = 2^a$ to order $16n$.

Consider a square, real STBC of order $n = 2^a$

$$\mathcal{G}_{2^a} = \mathcal{G}_{2^a}(x_1, \ldots, x_{\rho(n)})$$

which has $\rho(n) = \rho(2^a)$ real variables. From $\mathcal{G}_{2^a}$, we can construct a square, real STBC of order $16n = 2^{a+4}$ with $\rho(2^a) + 8$ real variables x_i for $i = 1, \ldots, \rho(2^a) + 8$, denoted by

$$\mathcal{G}_{2^{a+4}} = \mathcal{G}_{2^{a+4}}(x_1, \ldots, x_{\rho(2^a)+8})$$

as given below

$$\mathcal{G}_{2^{a+4}} = \begin{bmatrix} \mathbf{I}_n \otimes \mathbf{L}_4(y_1, y_2, y_3, y_4) & \mathbf{O}_{4n} \\ \mathbf{O}_{4n} & \mathbf{I}_n \otimes \mathbf{L}_4(y_1, y_2, y_3, y_4) \\ \mathbf{I}_n \otimes \mathbf{R}_4(y_5, -y_6, -y_7, -y_8) & \mathcal{G}_{2^a} \otimes \mathbf{I}_4 \\ \mathcal{G}_{2^a}^T \otimes \mathbf{I}_4 & \mathbf{I}_n \otimes \mathbf{R}_4(-y_5, -y_6, -y_7, -y_8) \end{bmatrix}$$

$$\begin{bmatrix} \mathbf{I}_n \otimes \mathbf{R}_4(y_5, y_6, y_7, y_8) & \mathcal{G}_{2^a} \otimes \mathbf{I}_4 \\ \mathcal{G}_{2^a}^T \otimes \mathbf{I}_4 & \mathbf{I}_n \otimes \mathbf{R}_4(-y_5, y_6, y_7, y_8) \\ \mathbf{I}_n \otimes \mathbf{L}_4(-y_1, y_2, y_3, y_4) & \mathbf{O}_{4n} \\ \mathbf{O}_{4n} & \mathbf{I}_n \otimes \mathbf{L}_4(-y_1, y_2, y_3, y_4) \end{bmatrix}$$

where $\mathbf{L}_4$ and $\mathbf{R}_4$ are defined as

$$\mathbf{L}_4(y_1, y_2, y_3, y_4) = \begin{bmatrix} y_1 & -y_2 & -y_3 & -y_4 \\ y_2 & y_1 & y_4 & -y_3 \\ y_3 & -y_4 & y_1 & y_2 \\ y_4 & y_3 & -y_2 & y_1 \end{bmatrix}$$

$$\mathbf{R}_4(y_5, y_6, y_7, y_8) = \begin{bmatrix} y_5 & -y_6 & -y_7 & -y_8 \\ y_6 & y_5 & -y_8 & y_7 \\ y_7 & y_8 & y_5 & -y_6 \\ y_8 & -y_7 & y_6 & y_5 \end{bmatrix}$$

while $\mathbf{O}_n$ is the zero matrix of order n and $y_i = x_{\rho(n)+i}$ for $i = 1, \ldots, 8$.

3 *The Geramita-Pullman construction [Geramita and Pullman, 1974]*: Assume that we are given a square, real STBC of order $n = 2^a$ with $\rho(n) = \rho(2^a)$ real variables, denoted by

$$\begin{aligned} \mathcal{G}_{2^a} &= \mathcal{G}_{2^a}(x_1, \ldots, x_{\rho(n)}) \\ &= x_1 \mathbf{I}_n + x_2 \mathbf{M}_2 + \cdots + x_{\rho(n)} \mathbf{M}_{\rho(n)} \end{aligned}$$

where $\mathbf{M}_i$ are the real coefficient matrices of order n for $i = 1, \ldots, \rho(n)$ ($\mathbf{M}_1 = \mathbf{I}_n$). The transition from order n to order $16n$ is presented as

$$\begin{aligned} &\mathcal{G}_{2^{a+4}}(x_1, \ldots, x_{\rho(2^a)+8}) \\ =\ & x_1 \mathbf{I}_{16n} \\ +\ & \sum_{i=2}^{\rho(n)} \begin{bmatrix} \mathbf{O}_{8n} & \mathbf{I}_8 \otimes x_i \mathbf{M}_i \\ \mathbf{I}_8 \otimes x_i \mathbf{M}_i & \mathbf{O}_{8n} \end{bmatrix} + \begin{bmatrix} \mathbf{O}_{8n} & x_{\rho(n)+1} \mathbf{I}_{8n} \\ -x_{\rho(n)+1} \mathbf{I}_{8n} & \mathbf{O}_{8n} \end{bmatrix} \\ +\ & \begin{bmatrix} \mathcal{G}_8(0, x_{\rho(n)+2}, \ldots, x_{\rho(n)+8}) \otimes \mathbf{I}_n & \mathbf{O}_{8n} \\ \mathbf{O}_{8n} & \mathcal{G}_8(0, -x_{\rho(n)+2}, \ldots, -x_{\rho(n)+8}) \otimes \mathbf{I}_n \end{bmatrix} \end{aligned}$$

In order to construct the square, real STBC of order $n = 2^a(2b + 1)$, from the square, real STBC $\mathcal{G}_{2^a}$ of order $n = 2^a$, we just need to perform the Kronecker product between $\mathbf{I}_{(2b+1)}$ and $\mathcal{G}_{2^a}$.

2.3.1.3 Constructions of Full-Rate, Non-Square, Real STBCs

The method to construct full-rate, non-square, real STBCs for any number n of Tx antennas is mentioned by Liang [Liang, 2003]. Readers may refer to the Part IV in [Liang, 2003] for more details. Again, it is noted that the *full-rate, square*, real STBCs exist only for $n=1, 2, 4$ and 8. For other values of n, the full-rate, real STBCs *must be non-square*.

2.3.2   Complex Orthogonal Designs - CODs

DEFINITION 2.3.5 *A generalized Complex, Orthogonal Design COD (also complex, orthogonal STBC)* $\mathbf{Z} = \mathbf{X} + i\mathbf{Y}$ *is defined as a* $p \times n$ *matrix whose nonzero entries are the indeterminates* $\pm s_1$, $\pm s_2$, ..., $\pm s_k$, *their conjugates* $\pm s_1^*$, $\pm s_2^*$, ..., $\pm s_k^*$ *or their products with* $i = \sqrt{-1}$ *over the complex number field* $\mathbb{C}$, *such that*

$$\mathbf{Z}^H \mathbf{Z} = \left(\sum_{j=1}^{k} |s_j|^2 \right) \mathbf{I}_{n \times n} \tag{2.29}$$

where $\mathbf{Z}^H$ *denotes the Hermitian transpose of* $\mathbf{Z}$ *and* $\mathbf{I}_{n \times n}$ *is the identity matrix of order* n. *The rate of* $\mathbf{Z}$ *is* $R = k/p$. *The matrix* $\mathbf{Z}$ *is said to be a* $[p, n, k]$ *complex orthogonal design. If p=n,* $\mathbf{Z}$ *is called square, complex orthogonal design (square COD). Otherwise,* $\mathbf{Z}$ *is called non-square (or rectangular) COD.*

It is important to note that the Complex Orthogonal Designs (CODs) defined as above are actually the so-called Generalized Complex Orthogonal Designs (GCODs) *without Linear Processing (LP)*, which were first used by Tarokh et al. [Tarokh et al., 1999b]. Meanwhile, the term "Complex Orthogonal Designs" (CODs) in [Tarokh et al., 1999b] was used to present a class of complex orthogonal designs which are *square*, of size $n \times n$ and comprise n indeterminates. However, nowadays, the term "CODs" is usually referred to as the one defined by the Definition 2.3.5.

DEFINITION 2.3.6 *A generalized Complex Orthogonal Design COD with Linear Processing - LP (also complex linear processing orthogonal STBC)* $\mathbf{Z} = \mathbf{X} + i\mathbf{Y}$ *is defined as a* $p \times n$ *matrix whose nonzero entries are the indeterminates* $\pm s_1$, $\pm s_2$, ..., $\pm s_k$, *their conjugates* $\pm s_1^*$, $\pm s_2^*$, ..., $\pm s_k^*$ *or their products with* $i = \sqrt{-1}$ *over the complex number field* $\mathbb{C}$, *such that*

$$\mathbf{Z}^H \mathbf{Z} = \mathcal{D}$$

where $\mathcal{D}$ *is a diagonal matrix of size* $n \times n$ *with diagonal entries* $\mathcal{D}_{jj}$, *for* $j = 1, 2, \ldots, n$, *of the form* $(l_{j1}|x_1|^2 + l_{j2}|x_2|^2 + \cdots + l_{jk}|x_k|^2)$. *The coefficients* $l_{j1}, \ldots, l_{jk}$ *are strictly positive, real numbers.* $\mathbf{Z}^H$ *denotes the Hermitian transpose of* $\mathbf{Z}$. *The rate of* $\mathbf{Z}$ *is* $R = k/p$. *The matrix* $\mathbf{Z}$ *is said to be a* $[p, n, k]$ *COD with LP. If p=n,* $\mathbf{Z}$ *is called square, complex orthogonal design (square COD) with LP. Otherwise,* $\mathbf{Z}$ *is called non-square (or rectangular) COD with LP.*

It is important to note that CODs with LP defined as above were originally called Generalized Complex Orthogonal Designs (GCODs) *with LP* by Tarokh et al. [Tarokh et al., 1999b].

By the similar analysis as mentioned in Section 2.3.1, it is easy to realize that the matrix $\mathbf{Z}$ defined in the Definitions 2.3.5 and 2.3.6 provides a full diversity order as $\det(\mathbf{Z}^H\mathbf{Z}) > 0$ for any distinct pair of codewords $\mathbf{s} \triangleq (s_1, \ldots, s_k)$ and $\acute{\mathbf{s}} \triangleq (\acute{s}_1, \ldots, \acute{s}_k)$, provided that the coefficients l_{jl}, for $j = 1, \ldots, n$ and $l = 1, \ldots, k$ are definitely positive.

As opposed to real STBCs, CODs (or Complex Orthogonal STBCs - CO STBCs) can be used for PSK and QAM modulations. Also, they have not been well known unlike real STBCs.

2.3.2.1 Maximum Rate of *Square, Complex Orthogonal* STBCs

It has been proved that the maximum number k_{max} of variables in a *square* COD of size $n = 2^a(2b+1)$ is [Adams et al., 1965], [Liang, 2003], [Tirkkonen and Hottinen, 2002]:

$$k_{max} = (a+1) \tag{2.30}$$

From Eq. (2.30), we have the following corollary:

COROLLARY 2.3.7 *The maximum rate of square, complex orthogonal designs for any number $n = 2^a(2b+1)$ of transmitter antennas is*

$$R_{Cmax} = \frac{k_{max}}{n} = \frac{a+1}{2^a(2b+1)} \tag{2.31}$$

In (2.31), the subscript "C" implies that the formula is applied to complex designs.

The maximum number of variables, k_{max}, and the maximum rates R_{Cmax} of *square, complex, orthogonal* STBCs (CO STBCs) for some typical values of n are given in Table 2.3.

REMARK 2.3.8 *It is important to clarify that, according to Liang's paper [Liang, 2003], the maximum achievable rate for CO STBCs of orders $n = 2m - 1$ or $n = 2m$ is (see Eq. (130) in [Liang, 2003])*

$$R_{max} = (m+1)/2m \tag{2.32}$$

However, note that this maximum rate is only achievable for non-square constructions, except for the special case when $m = 1$, i.e. when $n = 1$ or $n = 2$.

Table 2.3. The maximum number of variables and the maximum rates of *square* CO STBCs.

n	k_{max}	R_{Cmax}	n	k_{max}	R_{Cmax}
1	1	1	9	1	1/9
2	2	1	10	2	1/5
3	1	1/3	11	1	1/11
4	3	3/4	12	4	1/3
5	1	1/5	13	1	1/13
6	2	1/3	14	2	1/7
7	1	1/7	15	1	1/15
8	4	1/2	16	5	5/16

For square constructions of orders $n = 2^a(2b + 1)$, the maximum achievable rate must be calculated by Eq. (2.31). When $m = 1$, (2.31) and (2.32) provide the same results. Readers should refer to Corollary 2 and Section II D in [Liang, 2003], or Section IV in [Tirkkonen and Hottinen, 2002] for more details.

Particularly, for $n = 8$, i.e., $m = 4$, $a = 3$ and $b = 0$, the maximum achievable rate of non-square CO STBCs following (2.32) is 5/8, while the maximum achievable rate of square CO STBCs according to (2.31) is only 1/2. In Liang's paper, the author made an unclear statement in the abstract that the achievable maximum rate for $n = 2m - 1$ and $n = 2m$ is $(m + 1)/2m$, but did not state if this maximum rate is achievable by square or non-square constructions, which may lead to some confusion.

From Eq. (2.31) and Table 2.3, we have some following important notes:

1 If n is odd, the maximum code rate of *square* CO STBCs is $R_{Cmax} = 1/n$. For $n > 3$, the code rate is very small, and hence, there might be no point to examine *square* CO STBC for the odd number of Tx antennas which is greater than 3.

2 *Square*, CO STBCs with a full rate exist only for $n = 1$ and $n = 2$ Tx antennas.

2.3.2.2   Maximum Possible Rate of *Non-Square* CO STBCs

If n can be presented as $n = 2m - 1$ or $n = 2m$, where m is any nonzero natural number, then the maximum possible rate of *non-square* CO STBCs is given by the following theorem, which is equivalent to Theorems 5 and 6 in [Liang, 2003]:

THEOREM 2.3.9 *For any given number of transmitter antennas $n = 2m - 1$ and $n = 2m$ with $m \in \mathbb{N}$ and $m \neq 0$, the rate of non-square, complex orthogonal STBCs satisfies*

$$R \leq \frac{m+1}{2m} \tag{2.33}$$

From the above theorem, we can draw the following corollary:

COROLLARY 2.3.10 *For any number of Tx antennas $n = 2m-1$ and $n = 2m$ with $m \in \mathbb{N}$ and $m \neq 0$, there exist certain values of the two parameters k (the number of variables in the code) and p (the length of the code) for which the $[p, n, k]$ non-square CO STBC achieves the maximum rate:*

$$R_{max} = \frac{m+1}{2m} \tag{2.34}$$

EXAMPLE 2.3.1 *For $n = 8$, the maximum rate of non-square CO STBCs is R_{max}=5/8. As clarified earlier in Remark 2.3.8, this maximum rate does not contradict with the maximum rate R_{Cmax}=1/2 mentioned in Table 2.3 for the same n. It is because the maximum rate R_{max}=5/8 is for non-square orthogonal designs, while the maximum rate R_{Cmax}=1/2 is for square orthogonal designs. The maximum rate-5/8, non-square CO STBC exists for $n = 8$, $p = 112$, $k = 70$, i.e., [112,8,70] CO STBC. This construction can be found in Appendix E in [Liang, 2003].*

Some typical values of the maximum possible rates R_{max}, the maximum number of variables k_{max}, and the optimal delay p_{min} of *non-square* CO STBCs are given in Table 2.4, Table 2.5 and Table 2.6, respectively. These tables are derived from Table I and Table II in [Liang, 2003].

From these tables, some following important notes are derived:

1 If n is odd, as opposed to the case of *square* CO STBCs (see Table 2.3), the maximum code rate of *non-square* CO STBCs are still potentially high. For instance, when n=15, the maximum code rate in the former case is R_{Cmax}=1/15 while that in the later case is R_{max}=9/16.

2 Although having higher maximum rates, *non-square* CO STBCs require a very large decoding delay (also memory length) for $n > 6$ (see Table 2.6).

Table 2.4.   The maximum possible rates of *non-square* CO STBCs.

n	R_{max}	n	R_{max}
1	1	9	3/5
2	1	10	3/5
3	3/4	11	7/12
4	3/4	12	7/12
5	2/3	13	4/7
6	2/3	14	4/7
7	5/8	15	9/16
8	5/8	16	9/16

Table 2.5.   The maximum number of variables of *non-square* CO STBCs.

n	k_{max}	n	k_{max}
1	1	9	126
2	2	10	252
3	3	11	462
4	6	12	924
5	10	13	1716
6	20	14	3432
7	35	15	6435
8	70	16	12870

Table 2.6.   The optimal delay of *non-square* CO STBCs with the maximum possible rates.

n	p_{min}	n	p_{min}
1	1	9	210
2	2	10	420
3	4	11	792
4	8	12	1584
5	15	13	3003
6	30	14	6006
7	56	15	11440
8	112	16	22880

2.3.2.3   Constructions of Maximum Rate, *Square* CO STBCs

In this section, three methods for the construction of *maximum rate, square* CO STBCs, namely 1) the Jozefiak construction; 2) the Adams-Lax-Phillips construction and 3) the Wolfe construction, are presented. It is noted that there exist other construction methods. For instance, from Clifford representation theory, Tirkkonen et al. [Tirkkonen and Hottinen, 2002] proposed another method to construct *maximum rate, square* CO STBCs. Readers may refer to Section 3.1 in [Su and Xia, 2003] or Eq. (20) in [Tirkkonen and Hottinen, 2002] for more details. Other construction methods follow from the Amicable Orthogonal Designs (AODs), which are fully explained in [Geramita and Seberry, 1979] and will be mentioned in more details later in Chapter 3.

1 *The Jozefiak construction [Jozefiak, 1976]*: Assume that n is even and that $n = 2^a$. Further, assume that we already have an $n \times n$ square, complex orthogonal design $\mathbf{Z}_{2^a} = \mathbf{Z}_{2^a}(s_1, \ldots, s_{a+1})$. Then the square complex STBC of size $2n \times 2n$ is constructed as

$$
\begin{aligned}
\mathbf{Z}_{2^{a+1}} &= \mathbf{Z}_{2^{a+1}}(s_1, \ldots, s_{a+1}, s_{a+2}) \\
&= \begin{bmatrix} \mathbf{Z}_{2^a} & s_{a+2}\mathbf{I}_n \\ -s_{a+2}^*\mathbf{I}_n & \mathbf{Z}_{2^a}^H \end{bmatrix}
\end{aligned}
\tag{2.35}
$$

To construct the *maximum rate, square* CO STBC of size $n = 2^a(2b + 1)$, we need to perform the Kronecker product between the identity matrix $\mathbf{I}_{(2b+1)}$ and the maximum rate, square, complex STBC of size $n = 2^a$, i.e. $\mathbf{Z}_{2^a(2b+1)} = \mathbf{I}_{(2b+1)} \bigotimes \mathbf{Z}_{2^a}$.

EXAMPLE 2.3.2
For n=1, i.e., a=0, we have $\mathbf{Z}_1 = (s_1)$.
For n=2, i.e., a=1, from (2.35), we have

$$
\mathbf{Z}_2 = \begin{bmatrix} s_1 & s_2 \\ -s_2^* & s_1^* \end{bmatrix}
$$

For n=3, i.e., a=0, b=1, we have

$$
\mathbf{Z}_3 = \mathbf{I}_3 \bigotimes \mathbf{Z}_1 = \begin{bmatrix} s_1 & 0 & 0 \\ 0 & s_1 & 0 \\ 0 & 0 & s_1 \end{bmatrix}
$$

For n=4, i.e., a=2, we have

$$
\mathbf{Z}_4 = \begin{bmatrix} s_1 & s_2 & s_3 & 0 \\ -s_2^* & s_1^* & 0 & s_3 \\ -s_3^* & 0 & s_1^* & -s_2 \\ 0 & -s_3^* & s_2^* & s_1 \end{bmatrix}
$$

For n=8, i.e., a=3, we have

$$\mathbf{Z}_8 \;=\; \left[\begin{array}{c|c} \mathbf{Z}_4 & s_4\mathbf{I}_4 \\ \hline -s_4^*\mathbf{I}_4 & \mathbf{Z}_4^H \end{array}\right]$$

$$=\; \left[\begin{array}{cccc|cccc} s_1 & s_2 & s_3 & 0 & s_4 & 0 & 0 & 0 \\ -s_2^* & s_1^* & 0 & s_3 & 0 & s_4 & 0 & 0 \\ -s_3^* & 0 & s_1^* & -s_2 & 0 & 0 & s_4 & 0 \\ 0 & -s_3^* & s_2^* & s_1 & 0 & 0 & 0 & s_4 \\ \hline -s_4^* & 0 & 0 & 0 & s_1^* & -s_2 & -s_3 & 0 \\ 0 & -s_4^* & 0 & 0 & s_2^* & s_1 & 0 & -s_3 \\ 0 & 0 & -s_4^* & 0 & s_3^* & 0 & s_1 & s_2 \\ 0 & 0 & 0 & -s_4^* & 0 & s_3^* & -s_2^* & s_1^* \end{array}\right]$$

$$(2.36)$$

2 *The Adams-Lax-Phillips construction [Adams et al., 1965]*: This construction is similar to the Jozefiak construction except that the recursive formula (2.35) is replaced by

$$\begin{aligned} \mathbf{Z}_{2^{a+1}} &= \mathbf{Z}_{2^{a+1}}(s_1, \ldots, s_{a+1}, s_{a+2}) \\ &= \left[\begin{array}{cc} s_{a+2}\mathbf{I}_n & \mathbf{Z}_1 \\ \mathbf{Z}_1^H & -s_{a+2}^*\mathbf{I}_n \end{array}\right] \end{aligned}$$

3 *The Wolfe construction [Wolfe, 1976]*: This construction is similar to the Jozefiak construction and the Adams-Lax-Phillips construction except that the recursive formula (2.35) is replaced by

$$\begin{aligned} \mathbf{Z}_{2^{a+1}} &= \mathbf{Z}_{2^{a+1}}(s_1, \ldots, s_{a+1}, s_{a+2}) \\ &= \left[\begin{array}{cc} s_{a+2}\mathbf{I}_n & \mathbf{Z}_1 \\ -\mathbf{Z}_1^H & s_{a+2}^*\mathbf{I}_n \end{array}\right] \end{aligned}$$

2.3.2.4   Constructions of High-Rate, *Non-Square* CO STBCs

The method to construct high-rate, *non-square* CO STBCs for any number n of Tx antennas is mentioned by Liang [Liang, 2003]. Readers may refer to the Part V in [Liang, 2003] for more details. It should be emphasized that, by this method, maximum rate, *square* CO STBCs can also be constructed. However, the constructing procedures are more complicated than the methods mentioned in Section 2.3.2.3. Therefore, this construction method should only be used for constructing high-rate, *non-square* CO STBCs.

2.3.2.5   On the Maximum Rates of CODs with Linear Processing

CODs with Linear Processing (LP) are defined by Definition 2.3.6. It has been shown by the Corollary 4.1 in [Su and Xia, 2003] that, if the coefficients satisfy $l_{j1} = \cdots = l_{jk}$, for $j = 1, \ldots, n$, then relaxing the definition of CODs in order to allow LP at the transmitter fails to provide a higher code rate for *square* CODs (also *square* CO STBCs). In other words, the maximum code rate of *square* CO STBCs with LP is also calculated by Eq. (2.31).

However, it is *unknown* whether the maximum rate of *square* CO STBCs *with* LP is different to that of *square* CO STBCs *without* LP if the condition $l_{j1} = \cdots = l_{jk}$, for $j = 1, \ldots, n$, is *not* satisfied.

As opposed to *square* CO STBCs with LP, the maximum rate of *non-square* CO STBCs with LP has been *unknown* yet.

Since full-rate, *real* STBCs exist for any number of Tx antennas (see the second note of Section 2.3.1.1), we are always able to construct rate-1/2, *non-square* CO STBCs with LP for any number of Tx antennas. The construction method is mentioned by Part E in [Tarokh et al., 1999b] and by Eq. (4.4) in [Su and Xia, 2003]. Furthermore, in [Su and Xia, 2003] and [Su and Xia, 2001], W. Su et al. derived the two *non-square* CO STBCs *with LP* for 5 and 6 Tx antennas with code rates of 7/11 and 18/30 (or 0.6), respectively. At the time of their discovery, those codes were the *known maximum-rate, non-square* CO STBCs *with LP* for 5 and 6 Tx antennas.

However, up to date, these code rates were outdated already. By using the construction method proposed in [Liang, 2003] with the observation that the orthogonality is not affected by multiplying each column of a COD with a coefficient l ($l \neq 0$ and $l \neq 1$), we realize that the achievable code rate of *non-square* CO STBCs *with LP* is also calculated by Eq. (2.34). This is the *known maximum rate* of *non-square* CO STBCs *with LP* so far.

For instance, by multiplying the first column of each of the constructions (100) and (101) in [Liang, 2003] for 5 and 6 Tx antennas, i.e. the [15,5,10] and [30,6,20] CO STBCs, respectively, with a coefficient $l = 2$, we have the *non-square* CO STBCs *with LP* $\mathbf{Z}$ satisfying

$$\mathbf{Z}^H\mathbf{Z} = diag(4\sum_{i=1}^{10}|s_i|^2, \sum_{i=1}^{10}|s_i|^2, \sum_{i=1}^{10}|s_i|^2, \sum_{i=1}^{10}|s_i|^2, \sum_{i=1}^{10}|s_i|^2)$$

$$\mathbf{Z}^H\mathbf{Z} = diag(4\sum_{i=1}^{20}|s_i|^2, \sum_{i=1}^{20}|s_i|^2, \sum_{i=1}^{20}|s_i|^2, \sum_{i=1}^{20}|s_i|^2, \sum_{i=1}^{20}|s_i|^2, \sum_{i=1}^{20}|s_i|^2)$$

where *diag* denotes a diagonal matrix with the elements on the main diagonal provided in the brackets. The code rates of these *non-square* CO STBCs *with LP* are 2/3 and 2/3, respectively, which are higher than the code rates 7/11 and 18/30 in [Su and Xia, 2003], [Su and Xia, 2001].

Therefore, we can conclude that the *known maximum rate* of *non-square* CO STBCs *with LP* to date is the same as that of *non-square* CO STBCs *without LP* and is calculated by Eq. (2.34) for any number of Tx antennas. However, it is *unknown* whether or not the *true* maximum rate of *non-square* CO STBCs *with LP* is higher than that of *non-square* CO STBCs *without LP*. This requires to be further examined.

2.3.2.6   Capacity of the Channel Using STBCs

We consider here a CO STBC of size $p \times n_T$ comprising k complex variables, where n_T denotes the number of Tx antennas and p denotes the length of the CO STBC. The code rate is thus $R = k/p$. We assume that channels comprise n_R Rx antennas and they are *block*, flat Rayleigh fading channels. Channel coefficients are assumed to be *known at the receiver*, but *unknown at the transmitter*. Therefore, the transmitted power per Tx antenna is assumed to be the same and equal to P/n_T (see Section 2.2.3 for more details).

According to S. Sandhu et al. [Sandhu and Paulraj, 2000], although using STBCs can provide a high rate and a full transmission diversity order for a given number of Tx antennas with relatively simple detection, it incurs a loss of capacity in comparison with the true capacity of the channel.

The loss of capacity is a function of the rank r of the channel coefficient matrix $\mathbf{H}$ ($r \leq \min(n_T, n_R)$) and the code rate $R = k/p$. As a result, the loss of capacity depends on the number of Tx antennas n_T and Rx antennas n_R, and the code rate $R = k/p$ (see Eq. (6) in [Sandhu and Paulraj, 2000]):

$$
\begin{aligned}
\Delta C \;=\; & W \left(1 - \frac{k}{p} \right) E \left\{ \log_2 \left(1 + \frac{P}{n_T \sigma^2} \parallel \mathbf{H} \parallel_F^2 \right) \right\} \\
& + \; \log_2 \left(1 + \frac{S}{1 + \frac{P}{n_T \sigma^2} \parallel \mathbf{H} \parallel_F^2} \right)
\end{aligned}
\tag{2.37}
$$

where

$$
\parallel \mathbf{H} \parallel_F^2 = \sum_{i=1}^{r} \lambda_i
$$

$$S = \left(\frac{P}{n_T\sigma^2}\right)^2 \sum_{i1<i2}^{r} \lambda_{i1}\lambda_{i2} + \left(\frac{P}{n_T\sigma^2}\right)^3 \sum_{i1<i2<i3}^{r} \lambda_{i1}\lambda_{i2}\lambda_{i3} +$$

$$+ \ldots + \left(\frac{P}{n_T\sigma^2}\right)^r \prod_{i=1}^{r} \lambda_i$$

while λ_is are nonzero eigenvalues of $\mathbf{H}^H\mathbf{H}$ (or $\mathbf{H}\mathbf{H}^H$), W is the bandwidth of each sub-channel, $\frac{P}{\sigma^2}$ is the SNR at each Rx antenna. Readers may refer to Section 2.2 for more details on these notations.

From (2.37), some results of interest are derived as follows [Sandhu and Paulraj, 2000]:

- Any channel with a *full rate* CO STBC (i.e. $k = p$), such as the Alamouti code [Alamouti, 1998], used over a channel with *one Rx antenna* ($r = n_R = 1$) is always optimal with respect to capacity since $\Delta C = 0$.

- CO STBCs of *any rates*, including the full-rate codes, such as the Alamouti code, used over i.i.d. Rayleigh channel with *multiple Rx antennas*, i.e. $n_R \geq 2$, always incur a loss in capacity because ΔC is non-zero.

Therefore, although using STBCs in MIMO systems can provide a potentially high capacity and a full transmission diversity order for a given number of Tx antennas with a relatively simple decoding algorithm, there exists a loss of capacity compared to the true maximum capacity of the MIMO systems.

2.3.2.7   Examples on the Capacity of Channels and of CO STBCs

We consider a MIMO system with n_T Tx and n_R Rx antennas. The transmission model is thus

$$\mathbf{Y} = \sqrt{\frac{\rho}{n_T}}\mathbf{X}\mathbf{H} + \mathbf{N} \tag{2.38}$$

where $\mathbf{X} \in \mathbb{C}^{p \times n_T}$ denotes the matrix of complex transmitted signals during p symbol time slots (or p channel uses), $\mathbf{H} \in \mathbb{C}^{n_T \times n_R}$ denotes the channel coefficient matrix, and $\mathbf{N} \in \mathbb{C}^{p \times n_R}$ denotes the additive noise matrix. We assume that the entries of $\mathbf{H}$ and $\mathbf{X}$ are i.i.d. complex Gaussian random variables with the distribution $\mathcal{CN}(0, 1)$, which implies that

$$\begin{aligned} E\{tr(\mathbf{H^H H})\} &= n_T n_R \\ E\{tr(\mathbf{X^H X})\} &= n_T p \end{aligned} \tag{2.39}$$

The coefficient $\sqrt{\frac{\rho}{n_T}}$ in Eq. (2.38) ensures that ρ is the SNR at each Rx antenna during each symbol time slot (channel use), independently of the number n_T of Tx antennas.

We assume further that the channel is a flat, block Rayleigh fading channel whose channel coefficients are perfectly known at the receiver, but not at the transmitter. Let $C(\rho, n_T, n_R)$ be the *true* capacity of the channel between the Tx and Rx antennas, while $C_{STBC}(\rho, n_T, n_R)$ be the capacity of the channel where the STBC is utilized.

Similarly to that mentioned earlier in Eq. (2.8), the channel capacity is calculated as

$$C = E\left\{ W \, \log_2 \left[\det \left(\mathbf{I}_r + \frac{P}{n_T \sigma^2} \mathbf{Q} \right) \right] \right\} \tag{2.40}$$

where r is the rank of the matrix $\mathbf{H}$ and $\mathbf{Q}$ is defined as

$$\mathbf{Q} = \begin{cases} \mathbf{H}\mathbf{H}^H & \text{if } n_T < n_R \\ \mathbf{H}^H\mathbf{H} & \text{if } n_T \geq n_R \end{cases} \tag{2.41}$$

Note that the matrix $\mathbf{Q}$ in Eq. (2.41) is defined by a formula which is slightly different to that mentioned in Eq. (2.9) in Section 2.2.3.1. That is because, in Eq. (2.38), $\mathbf{H}$ has size $n_T \times n_R$, rather than having size $n_R \times n_T$ like in Section 2.2.3.1. In the case where $\mathbf{H}$ has size $n_R \times n_T$, we just need to exchange n_T and n_R in Eq. (2.41).

The channel capacity can be easily calculated (see Section 2.2.3.1 for more details). From Eq. (2.10), (2.11), (2.13) and (2.15), the channel capacity for some particular values of n_T, n_R and ρ is shown in Table 2.7.

We may have a question of how well an STBC performs from capacity perspective when comparing the maximum mutual information which can be supported to the true channel capacity. This question has been partially answered in Section 2.3.2.6. In this section, we calculate the maximum mutual information of some known STBCs in Rayleigh fading channels to expose more clearly this issue.

EXAMPLE 2.3.3 *We consider here a system with $n_T = 2$ and $n_R = 1$ using the Alamouti code*

$$\mathbf{X} = \begin{bmatrix} x_1 & x_2 \\ -x_2^* & x_1^* \end{bmatrix}$$

Table 2.7. Normalized channel capacity for several values of transmitter and receiver antenna numbers.

n_T	n_R	ρ (dB)	Formula for $\frac{C(\rho,n_T,n_R)}{W}$	$\frac{C(\rho,n_T,n_R)}{W}$
2	1	20	$\int_0^\infty \log_2\left(1+\frac{\rho\lambda}{n_T}\right)\lambda e^{-\lambda}d\lambda$	6.2810
2	2	20	$\int_0^\infty \log_2\left(1+\frac{\rho\lambda}{n_T}\right)[1+(1-\lambda)^2]e^{-\lambda}d\lambda$	11.2898
3	1	20	$\int_0^\infty \frac{1}{2!}\log_2\left(1+\frac{\rho\lambda}{n_T}\right)\lambda^2 e^{-\lambda}d\lambda$	6.4115
		21.25		6.8213
3	2	20	$\int_0^\infty \log_2\left(1+\frac{\rho\lambda}{n_T}\right)[1+\frac{1}{2}(2-\lambda)^2]\lambda e^{-\lambda}d\lambda$	12.1396
4	1	20	$\int_0^\infty \frac{1}{3!}\log_2\left(1+\frac{\rho\lambda}{n_T}\right)\lambda^3 e^{-\lambda}d\lambda$	6.4751
		23		7.4656
4	2	20	$\int_0^\infty \log_2\left(1+\frac{\rho\lambda}{n_T}\right)[\frac{1}{2!}+\frac{1}{6}(3-\lambda)^2]\lambda^2 e^{-\lambda}d\lambda$	12.4875
8	4	20	$\int_0^\infty \log_2\left(1+\frac{\rho\lambda}{n_T}\right)[\frac{1}{4!}+\frac{1}{5!}(5-\lambda)^2$ $+\frac{2}{6!}(\frac{\lambda^2}{2}-6\lambda+15)^2$ $+\frac{1}{7!12}(-\lambda^3+21\lambda^2-126\lambda+210)^2]\lambda^4 e^{-\lambda}d\lambda$	24.9326

From (2.38), the transmission model becomes

$$\begin{bmatrix} y_1 \\ y_2 \end{bmatrix} = \sqrt{\frac{\rho}{2}}\begin{bmatrix} x_1 & x_2 \\ -x_2^* & x_1^* \end{bmatrix}\begin{bmatrix} h_1 \\ h_2 \end{bmatrix}+\begin{bmatrix} n_1 \\ n_2 \end{bmatrix} \qquad (2.42)$$

We rewrite it as

$$\begin{bmatrix} y_1 \\ y_2^* \end{bmatrix} = \sqrt{\frac{\rho}{2}}\begin{bmatrix} h_1 & h_2 \\ h_2^* & -h_1^* \end{bmatrix}\begin{bmatrix} x_1 \\ x_2 \end{bmatrix}+\begin{bmatrix} n_1 \\ n_2^* \end{bmatrix}$$

Therefore, we have a modified transmission model as follows:

$$\acute{\mathbf{Y}} = \sqrt{\frac{\rho}{2}}\acute{\mathbf{H}}\acute{\mathbf{X}} + \acute{\mathbf{N}} \qquad (2.43)$$

It is easy to realize that the channel matrix $\mathbf{H}$ in (2.42) is changed into the channel matrix $\acute{\mathbf{H}}$, which is orthogonal and has a rank $r = 2$. From (2.40) and (2.43), the maximum mutual information of the Alamouti code per symbol time *slot (or per channel user - PCU) is calculated as*

$$
\begin{aligned}
C_{STBC}(\rho, 2, 1) &= \frac{1}{p} E\left\{ W \log_2 \left[\det\left(\mathbf{I}_r + \frac{\rho}{n_T} \acute{\mathbf{H}}^H \acute{\mathbf{H}} \right) \right] \right\} \\
&= \frac{1}{2} E\left\{ W \log_2 \left[\det\left(\mathbf{I}_2 + \frac{\rho}{2} \acute{\mathbf{H}}^H \acute{\mathbf{H}} \right) \right] \right\} \\
&= \frac{1}{2} E\left\{ W \log_2 \left[1 + \frac{\rho}{2}(|h_1|^2 + |h_2|^2) \right]^2 \right\} \\
&= E\left\{ W \log_2 \left[1 + \frac{\rho}{2}\left(|h_1|^2 + |h_2|^2 \right) \right] \right\}
\end{aligned}
$$

We see that $C_{STBC}(\rho, 2, 1)/W \equiv C(\rho, 2, 1)/W$, where $C(\rho, 2, 1)$ is the capacity of a MIMO system comprising two Tx and one Rx antennas with the SNR ρ at the Rx antenna (see (2.8)). Therefore, $C_{STBC}(\rho, 2, 1)/W = 6.2810$ bits/s/Hz per channel use (PCU) for $\rho = 20dB$ (see the Table 2.7). In other words, the Alamouti code in the channel with only one Rx antenna does not incur a loss of capacity in comparison with the true capacity of the channel. This agrees with the note in Section 2.3.2.6.

EXAMPLE 2.3.4 *We now show that the Alamouti code used in the channel with more than one Rx antenna does incur a loss of capacity. Assume that the channel now comprises $n_R = 2$ Rx antennas. The transmission model now becomes*

$$
\begin{bmatrix} y_{11} & y_{12} \\ y_{21} & y_{22} \end{bmatrix} = \sqrt{\frac{\rho}{2}} \begin{bmatrix} x_1 & x_2 \\ x_2^* & x_1^* \end{bmatrix} \begin{bmatrix} h_{11} & h_{12} \\ h_{21} & h_{22} \end{bmatrix} + \begin{bmatrix} n_{11} & n_{12} \\ n_{21} & n_{22} \end{bmatrix}
$$

It is rewritten as

$$
\begin{bmatrix} y_{11} \\ y_{21}^* \\ y_{12} \\ y_{22}^* \end{bmatrix} = \sqrt{\frac{\rho}{2}} \begin{bmatrix} h_{11} & h_{21} \\ h_{21}^* & -h_{11}^* \\ h_{12} & h_{22} \\ h_{22}^* & -h_{12}^* \end{bmatrix} \begin{bmatrix} x_1 \\ x_2 \end{bmatrix} + \begin{bmatrix} n_{11} \\ n_{21}^* \\ n_{12} \\ n_{22}^* \end{bmatrix}
$$

Similarly, the matrix $\acute{\mathbf{H}}$ *is orthogonal and has a rank* $r = 2$. *The maximum mutual information per channel use of the code is*

$$
\begin{aligned}
& C_{STBC}(\rho, 2, 2) \\
= {} & \frac{1}{p} E\left\{ W \log_2 \left[\det\left(\mathbf{I}_r + \frac{\rho}{n_T} \acute{\mathbf{H}}^H \acute{\mathbf{H}} \right) \right] \right\} \\
= {} & \frac{1}{2} E\left\{ W \log_2 \left[\det\left(\mathbf{I}_2 + \frac{\rho}{2}\left(|h_{11}|^2 + |h_{12}|^2 + |h_{21}|^2 + |h_{22}|^2 \right) \mathbf{I}_2 \right) \right] \right\} \\
= {} & E\left\{ W \log_2 \left[1 + \frac{2\rho}{4}\left(|h_{11}|^2 + |h_{12}|^2 + |h_{21}|^2 + |h_{22}|^2 \right) \right] \right\}
\end{aligned}
\tag{2.44}
$$

Therefore $C_{STBC}(\rho, 2, 2) \equiv C(2\rho, 4, 1)$, *where* $C(2\rho, 4, 1)$ *is the capacity of a MIMO system comprising 4 Tx antennas and 1 Rx antenna with the SNR* 2ρ *at the Rx antenna. Therefore,* $C_{STBC}(\rho, 2, 2) = 7.4656$ *bits/s/Hz PCU for* $\rho = 20dB$ *(see Table 2.7 for* $2\rho = 23$ *dB). The channel capacity in this case, meanwhile, is*

$$
\begin{aligned}
& C(\rho, 2, 2) \\
= {} & E\left\{ W \log_2 \left[\det\left(\mathbf{I}_r + \frac{\rho}{n_T} \acute{\mathbf{H}}\acute{\mathbf{H}}^H \right) \right] \right\} \\
= {} & E\left\{ W \log_2 \left[\left(1 + \frac{\rho}{2}(|h_{11}|^2 + |h_{12}|^2) \right)\left(1 + \frac{\rho}{2}(|h_{21}|^2 + |h_{22}|^2) \right) \right] \right\}
\end{aligned}
\tag{2.45}
$$

From (2.44) and (2.45), it is easy to prove that $C(\rho, 2, 2) > C_{STBC}(\rho, 2, 2)$. *In fact,* $C_{STBC}(\rho, 2, 2)/W = 7.4656$ *bits/s/Hz PCU while the normalized channel capacity* $C(\rho, 2, 2)/W = 11.2898$ *bits/s/Hz (see Table 2.7). Therefore, the Alamouti code in this case incurs a crucial loss of capacity. This also agrees with the note in Section 2.3.2.6.*

EXAMPLE 2.3.5 *We consider the scenario where* $n_T = 3$, $n_R = 1$, $p = 4$ *and the following non-square STBC (see [Hassibi and Hochwald, 2002], pp.1807):*

$$
\mathbf{X} = \sqrt{\frac{4}{3}} \begin{bmatrix} x_1 & x_2 & x_3 \\ -x_2^* & x_1^* & 0 \\ -x_3^* & 0 & x_1^* \\ 0 & -x_3^* & x_2^* \end{bmatrix}
$$

The coefficient $\sqrt{\frac{4}{3}}$ is to ensure the constraint (2.39). We have the equivalent channel coefficient matrix as given below:

$$\acute{\mathbf{H}} = \sqrt{\frac{4}{3}} \begin{bmatrix} h_1 & h_2 & h_3 \\ h_2^* & -h_1^* & 0 \\ h_3^* & 0 & -h_1^* \\ 0 & h_3^* & -h_2^* \end{bmatrix}$$

Therefore, the maximum mutual information per channel use of this STBC is

$$
\begin{aligned}
C_{STBC}(\rho, 3, 1) &= \frac{1}{p} E \left\{ W \log_2 \left[\det \left(\mathbf{I}_r + \frac{\rho}{n_T} \acute{\mathbf{H}}^H \acute{\mathbf{H}} \right) \right] \right\} \\
&= \frac{1}{4} E \left\{ W \log_2 \left[\det \left(\mathbf{I}_3 + \frac{\rho}{3} \frac{4}{3} \sum_{i=1}^{3} |h_i|^2 \mathbf{I}_3 \right) \right] \right\} \\
&= \frac{3}{4} E \left\{ W \log_2 \left(1 + \frac{\frac{4}{3}\rho}{3} \sum_{i=1}^{3} |h_i|^2 \right) \right\} \\
&= \frac{3}{4} C \left(\frac{4}{3} \rho, 3, 1 \right)
\end{aligned}
$$

Following Table 2.7, $C_{STBC}(\rho, 3, 1)/W = \frac{3}{4} C(\frac{4}{3}\rho, 3, 1)/W = 0.75 \times 6.8213 = 5.1160$ bits/s/Hz PCU for $\rho = 20dB$. Hence, the maximum mutual information $C_{STBC}(\rho, 3, 1)$ is smaller than the true channel capacity $C(\rho, 3, 1) = 6.4115$ bits/s/Hz.

Note that, it is not always possible to calculate the maximum mutual information of STBCs following the above method, since we cannot always find the equivalent channel coefficient matrix $\acute{\mathbf{H}}$ following the above method. The general method to find the equivalent channel coefficient matrix $\acute{\mathbf{H}}$ for calculating the maximum mutual information is mentioned in [Hassibi and Hochwald, 2002] (Section III). This method was originally derived for linear space-time codes, but is also applicable to STBCs.

2.4   Transmission Diversity Techniques

2.4.1   Classification of Transmission Diversity Techniques

Transmission diversity techniques have been widely used to enhance the performance of wireless channels. Various techniques have been proposed in the literature as well as applied in practice. These techniques can be classified as follows:

Depending on the domain in which the transmission redundancy is provided, diversity techniques are divided into *time diversity*, *frequency diversity* and *space diversity*.

Depending on where diversity techniques are used, transmission diversity techniques can be classified into *transmit diversity* and *receive diversity*.

2.4.1.1   Time Diversity

Time diversity is a diversity technique where identical signals are transmitted during different time slots. These time slots are uncorrelated, i.e., the temporal separation between those slots is greater than the *coherence time* of the wireless channel, which is in turn calculated as $1/F_m = c/(vF_s)$, where F_m is the maximum Doppler frequency, c the speed of light, v the speed of mobile and F_s the frequency of the transmitted signals [Proakis, 2001]. In fact, interleavers and error control coding, such as Forward Error Correction (FEC) codes, are employed to provide time diversity for the receiver. Another example of the modern implementation of time diversity is the RAKE receiver in CDMA (Code Division Multiple Access) systems [Rappaport, 2002] (pp. 391).

The main shortcoming of this technique is that the redundancy is provided in the time domain with a penalty of a loss in bandwidth efficiency. The loss in bandwidth is due to the guard time existing between the time slots.

2.4.1.2   Frequency Diversity

In this diversity technique, several frequencies are used to transmit the same signals. The frequency separation between these carrier frequencies is an order of several times of the *coherence bandwidth* of the channel [Proakis, 2001]. Consequently, the carrier frequencies are uncorrelated, i.e., they do not experience the same fades.

In practice, frequency diversity is often used in Line-Of-Sight (LOS) microwave channels. Some examples of systems employing frequency diversity include spread spectrum systems, such as Direct Sequence Spread Spectrum (DS-SS), Frequency Hoping Spread Spectrum (FH-SS) or Multi-Carrier Spread Spectrum (MC-SS) systems.

Similarly to the case of time diversity, in frequency diversity, the redundancy is provided in the frequency domain with the penalty of a loss in spectral efficiency. The loss in spectral efficiency is due to the guard bands existing between the carrier frequencies. Additionally, the structure of the receiver is complicated as it must be able to work with a number of frequencies.

2.4.1.3   Space Diversity

Space diversity, which is also named *antenna diversity*, has been frequently implemented in practice. This diversity technique can be further classified into various schemes, such as *polarization diversity* [Rappaport, 2002] (pp. 387), *beamforming diversity* [Blogh and Hanzo, 2002], [Godara, 1997], [Katz and Ylitalo, 2000], [Larsson et al., 2003], *antenna switching* [Barrett and Arnott, 1994]. Depending on whether it is applied to the transmitter or to the receiver, it can be classified into *transmit diversity* and *receive diversity*. Depending on how the replicas of the transmitted signals are combined at the receiver, *space diversity* techniques are classified into *selection combining* technique, *switched combining* technique, *equal-gain combining* technique and *maximum ratio combining* (MRC) technique [Rappaport, 2002].

The concept of space diversity is using multiple Tx and/or Rx antennas to transmit and/or receive signals. These antennas are spatially separated from one another by some halves of the wavelength, and consequently, they may be considered to be independent of one another [Jakes, 1974], [Salz and Winters, 1994].

Unlike time diversity and frequency diversity, in space diversity, the redundancy is provided for the receiver in the spatial domain, and consequently, this technique has no loss in spectral efficiency.

In practical wireless communication, a combined diversity technique of the aforementioned techniques is employed to provide multi-dimensional diversity. For instance, in GSM cellular systems, a combination between multiple antennas at the base station (space diversity) and interleaving as well as error control coding (time diversity) is utilized to provide the 2-dimensional diversity for the receivers (mobile users).

In this book, the authors examine the combination between multiple antennas (space diversity), antenna switching techniques (space diversity), Space-Time Block Codes STBCs (space and time diversity), and maximum ratio combining (MRC) technique (space diversity) to provide more diversity for wireless communication channels utilizing STBCs.

2.4.2   Spatial Diversity Combining Methods

As mentioned earlier, spatial diversity combining methods can be divided into the following categories: selection combining, scanning combining, maximum ratio combining and equal gain combining (see [Vucetic and Yuan, 2003] (pp. 55) and [Rappaport, 2002] (pp. 385)).

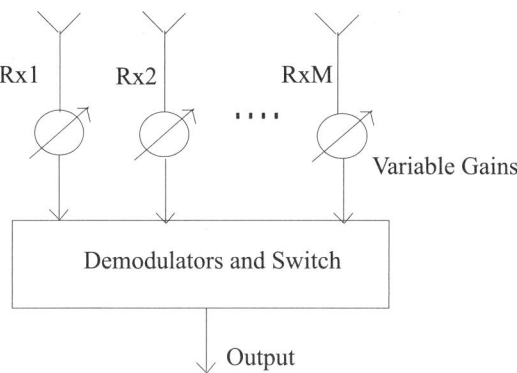

Figure 2.6. Selection combining method.

2.4.2.1 Selection Combining

This is the simplest spatial diversity combining method which requires only a SNR monitoring action and an antenna switch at the receiver.

In this technique, the receiver comprises M Rx antennas associated with M individual demodulators to provide M branches of which the gains are weighted to provide the same *average* SNR for every branch. The receiver selects the incoming signal with the highest *instantaneous* SNR to demodulate. In reality, as the measurement of the *instantaneous* SNR is difficult, the term $(S+N_0)/N_0$, where $(S + N_0)$ is the instantaneous power of the received signal (including noise), is normally measured. The diagram of the selection combining method is presented in Fig. 2.6.

The average SNR of the received signal $\bar{\gamma}$ (when selection combining is used) compared to the average SNR Γ of each branch (when no diversity is used) is calculated as follows (Eq. (7.62) in [Rappaport, 2002]):

$$\frac{\bar{\gamma}}{\Gamma} = \sum_{k=1}^{M} \frac{1}{k} \qquad (2.46)$$

Clearly, for $M \geq 2$, we have $\bar{\gamma} > \Gamma$. However, this technique is not optimal as it does not use all incoming signals simultaneously to provide the best received signals.

2.4.2.2 Scanning Combining - SC

In this technique, the receiver scans all branches following a certain order and selects a particular branch which has an SNR above the predetermined SNR threshold. The signal of this branch is selected as the output until it drops

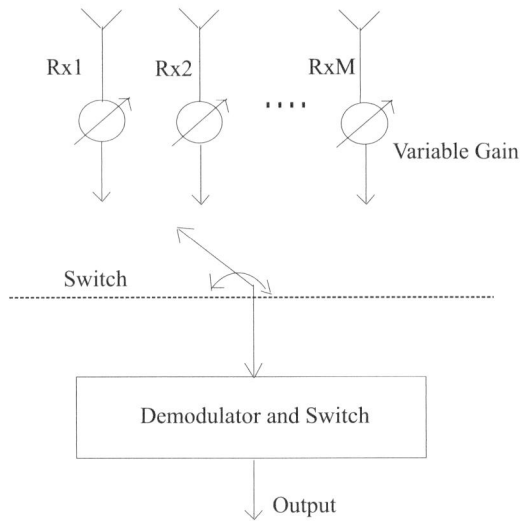

Figure 2.7.   Scanning combining method.

under the predetermined threshold. The receiver then starts searching again. The diagram of the scanning combining method is presented in Fig. 2.7.

The advantage of this technique over the selection combining method is that the receiver does not need to monitor continuous and instantaneous SNRs of all branches at all times. However, it is inferior compared to selection combining method as the best incoming signal is not always selected, and consequently, the average SNR of the output is smaller than that mentioned in (2.46).

2.4.2.3   Maximum Ratio Combining - MRC

In this technique, the signals from M branches are weighted by the corresponding weighting factors $G_k = \frac{r_k}{N_0}$ of those branches ($k = 1, \ldots, M$), where r_k and N_0 are the envelopes of received signals and the noise power, and then summed. Generally speaking, the weighting factors of the branches are *proportional* to the ratios $\frac{r_k}{N_0}$ of those branches themselves. Before summing, the signals must be co-phased to provide the coherence voltage addition. The average SNR $\bar{\gamma}_M$ of the output signal is simply the sum of individual SNRs Γ of all branches (Eq. (7.70) in [Rappaport, 2002]):

$$\bar{\gamma}_M = M\Gamma \tag{2.47}$$

Clearly, this technique can provide an acceptable output signal with the expected SNR even when no incoming signal is acceptable. Certainly, the structure and the cost of the MRC are higher than those of other combining methods. The

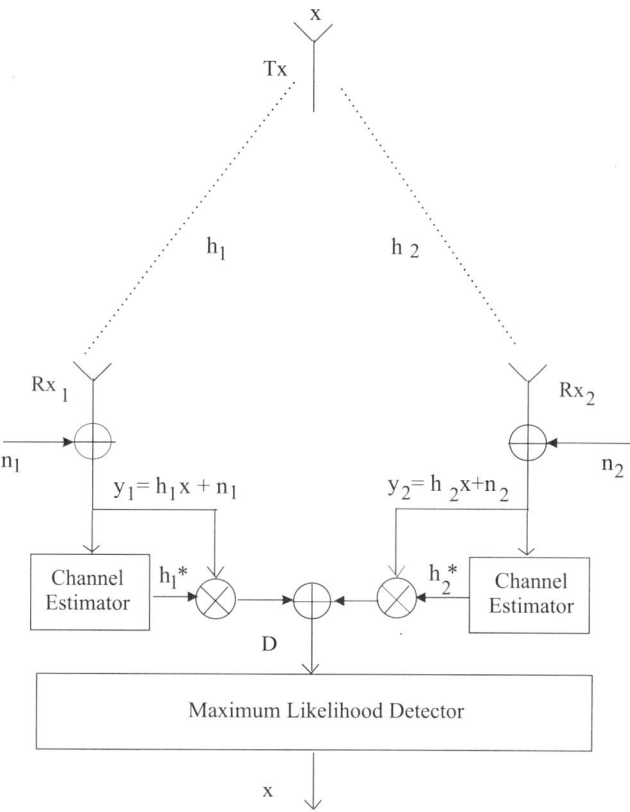

Figure 2.8. Conventional baseband MRC technique using two receiver antennas.

diagram of the conventional baseband MRC technique using 2 Rx antennas is presented in Fig. 2.8. Readers may refer to [Alamouti, 1998] or [Liew and Hanzo, 2002] (pp. 192) for more details.

2.4.2.4 Equal Gain Combining - EGC

This diversity technique is similar to the maximum ratio combining method, except that all weighting factors are set to one. The performance is marginally inferior compared to that of the maximum ratio combining method.

2.4.3 Transmit Diversity Techniques

Transmit diversity techniques [Fragouli et al., 2002], [Winters, 1994], [Winters, 1998] can be categorized to transmit diversity *with* or *without* feedback. In the uplinks (from Mobile Stations (MSs) to Base Stations (BSs)) of mobile communication systems, *receive diversity* is usually used by employing multiple receiver antennas at BSs. However, in the downlinks (from BSs to MSs), it is more practical to utilize *transmit diversity* techniques than using receive diversity techniques due to the following reasons:

- Multiple Rx antennas at MSs require a more complicated structure and more processing procedures at MSs, and consequently, require more power. Therefore, the lifetime of batteries in MSs is shortened.

- Due to the tiny size of MSs, it is impractical to install more than two Rx antennas at MSs.

Various transmit diversity techniques have been proposed in the literature, such as delay diversity schemes [Seshadri and Winters, 1993], [Wittneben, 1991], [Wittneben, 1993], beamforming [Larsson et al., 2003], antenna switching [Barrett and Arnott, 1994]. In order to improve further the performance of systems, transmit diversity is combined with *modulation schemes* and/or *coding schemes* with the consequence that both *diversity gain* and *coding gain* can be improved. The combination between transmit diversity techniques and Space-Time Codes (STCs) having coding gains, such as Space-Time Trellis Codes (STTCs) [Tarokh et al., 1998], is one of such combined transmit diversity techniques.

Although STBCs do not provide coding gains, STBCs possess a simple decoding method. They provide full diversity in both space and time domains *without* [1] or with a *small bandwidth expansion* [2]. The small bandwidth extension is due to the fact that the full-rate, CO STBCs do not exist for more than two Tx antennas (see Section 2.3.5 or [Liang, 2003], [Liang and Xia, 2003], [Wang and Xia, 2003]).

In this book, we propose some diversity Antenna Selection Techniques (ASTs) for channels using either STBCs or *differential* STBCs (DSTBCs). When STBCs or DSTBCs are used, it is possible to use the MRC technique at the receiver. The association between the proposed ASTs and the MRC technique significantly improves the performance of wireless channels using either STBCs or DSTBCs. Certainly, the performance of systems can be further improved by associating *transmit diversity* and *receive diversity*.

---

[1] For instance, when the Alamouti code [Alamouti, 1998] is used in systems with one receiver antenna. Readers may refer to Section 2.3.2.6 and 2.3.2.7 for more details.
[2] For the Alamouti code used in systems with more than one receiver antenna and for any CO STBC of order greater than 2. Readers may refer to Section 2.3.2.6 and 2.3.2.7 for more details.

2.5   Issues Addressed in the Book

As analyzed before, MIMO systems potentially possess a high capacity, which is a desired property for the current communication needs requiring a very high data rate and high reliability, such as multimedia communication services, cellular mobile, and the Internet. In many cases, as pointed out previously, the capacity of MIMO systems is approximately linearly proportional to the number of antennas.

STBCs are among various practical, advanced coding techniques designed for the use of multiple transmission antennas, which potentially approach the high capacity of MIMO systems (although there exists a loss of capacity compared to the true capacity of MIMO channels). As mentioned earlier, this coding technique possesses the following properties:

1 It completely or highly takes advantage of the channel bandwidth, i.e., the maximum mutual information of STBCs is equal (in the case of the Alamouti code used in systems with 1 Rx antenna) or approaches the high capacity of MIMO channels.

2 It decreases the sensitivity to multi-path fading, and facilitates utilization of higher level modulation schemes resulting in an increase of the data rate.

3 It also facilitates the increase of the coverage area of wireless systems, and consequently, improves the reuse of frequencies in mobile communication systems.

4 Consequently, it improves the data rate, the error performance and capacity of wireless communication channels with small or even without any expansion of bandwidth. An example is given in Fig. 2.9. This figure illustrates the bit error performance improvement of the Alamouti code compared to the transmission without space-time coding.

5 It requires a relatively simple decoding technique, i.e., Maximum Likelihood (ML) decoding, due to the orthogonality between the columns of the code matrices. Therefore, Space-Time Block coding is a simple and cost-effective coding scheme to meet the requirement on quality and spectral efficiency for the next generation wireless systems without changing much the structure of the existing systems. In fact, the Alamouti code has been adopted in the third (3G) generation wireless standards, such as WCDMA and CDMA 2000, by the Third Generation Partnership Project (3GPP) [3GPP, 2002], [Al-Dhahir, 2002], [Rappaport et al., 2002].

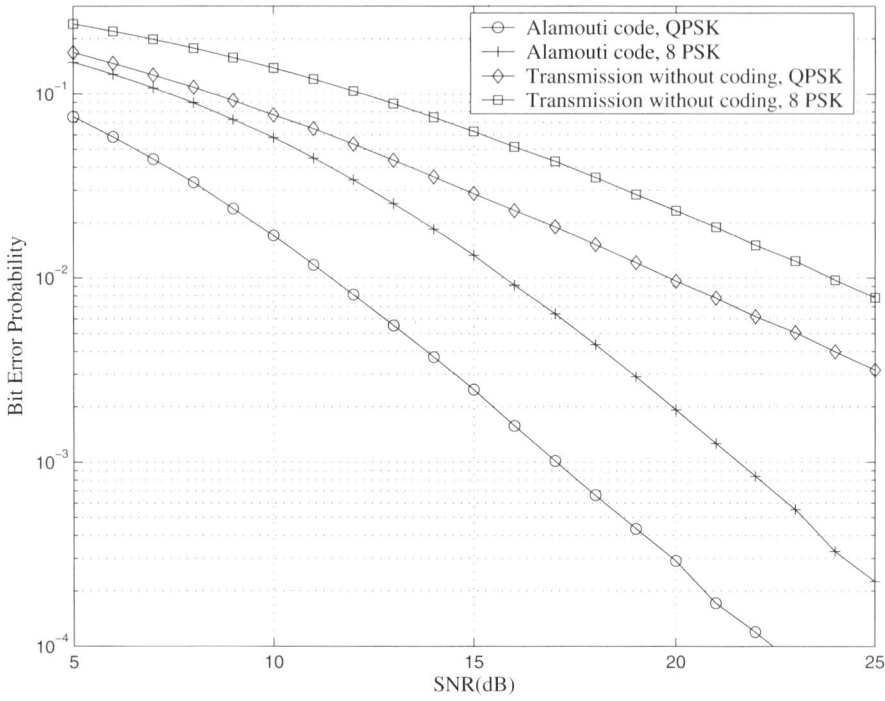

Figure 2.9. Alamouti code vs. transmission without coding with QPSK modulation and 8PSK modulation.

In this book, we consider the following research problems:

1 Although constructions of *square, maximum rate* CO STBCs, such as the Adams-Lax-Phillips construction, Jozefiak construction, and Wolfe construction, are well known (see Section 2.3.2.3 or [Liang, 2003]), the codes resulting from these constructions have numerous zeros, since these construction methods always involve identity matrices. This shortcoming impedes the practical implementation, especially in high data rate systems.

So far, the general constructions of *square, maximum rate* CO STBCs with fewer or even with no zeros such that the transmitted symbols equally disperse through Tx antennas have not been well examined. Those codes, referred to as the *improved square CO STBCs*, have the advantages that the power tends to be equally transmitted via each Tx antenna during every symbol time slot and that a lower peak power per Tx antenna is required to achieve the same bit error rates as for the conventional CO STBCs with zeros. *This book proposes such constructions of improved CO STBCs.*

2 To increase the bit rates of CO STBCs, Multi-Modulation Schemes (MMSs) can be applied to CO STBCs. The CO STBCs resulted from our aforementioned, discovered constructions, where some variables appear more often than the others, are especially good constructions for the use of MMSs. With the same MMSs and the same peak power per Tx antenna, such constructions may provide a *higher* data transmission rate and possess *better* error performance than the conventional STBCs where numerous zeros are present.

In addition, in MMSs, the optimal inter-symbol power allocation to achieve the best error performance is an important property. Although MMSs and the method examining the optimal inter-symbol power allocation have been somewhat mentioned in [Tirkkonen and Hottinen, 2001], these issues have not been well examined yet.

Therefore, this book proposes MMSs applied to CO STBCs to increase the transmission rate and a general method for examining the optimal inter-symbol power allocation in such MMSs to achieve the optimal error properties.

3 As analyzed earlier, the benefit of diversity techniques is evident. The combination between CO STBCs and a closed loop transmission diversity technique (using a feedback loop) to improve further the performance of wireless channels has been intensively examined in the case of *coherent detection*. However, it will be better if the time required to process the feedback information at the transmitter is shortened, and consequently, the system is quickly updated to the change of channels. *This book derives an improved transmitter antenna selection technique, which reduces the time required to process feedback information and enhances further the performance of space-time coded wireless channels in the case of coherent detection.*

4 So far, the closed loop transmit diversity techniques with the aid of a feedback loop and with *a limited number of training symbols (or pilot symbols)* to select transmitter and/or receiver antennas in the channels using Differential Space-Time Block Codes (DSTBCs) have not been intensively investigated yet. *This book proposes two transmission antenna selection techniques for the channels using DSTBCs, improving significantly the performance of*

wireless channels and being very robust, even in correlated, flat Rayleigh fading channels or in the channels with imperfect carrier recovery.

5   Although the methods to generate correlated Rayleigh fading envelopes in wireless channels have been intensely examined in the literature, those methods have their own shortcomings, which seriously limit their applicability. A more generalized algorithm to generate Rayleigh fading envelopes, which are correlated in *spatial*, *temporal* and/or *spectral* domains, and can be applied to the case of either *discrete-time instants* or *a real-time scenario*, is essential for researchers in simulating and modelling the channels. *This book proposes such a more generalized algorithm to generate correlated Rayleigh fading envelopes that overcomes all shortcomings of the conventional methods.*

Chapter 3

NEW SQUARE, COMPLEX ORTHOGONAL SPACE-TIME BLOCK CODES FOR EIGHT TRANSMITTER ANTENNAS

3.1 Introduction

Square, Complex Orthogonal Space-Time Block Codes (CO STBCs) are known for the relatively simple receiver structure and minimum processing delay in the case of complex signal constellations. One of the methods to construct square CO STBCs is based on Amicable Orthogonal Designs (AODs). The background knowledge on AODs can be found in [Geramita and Seberry, 1979]. The simplest CO STBC is the Alamouti code [Alamouti, 1998] for two transmitter (Tx) antennas, which is based on an amicable orthogonal pair of order-2 matrices. The Alamouti code achieves the transmission rate of one for 2 Tx antennas, while the CO STBCs for more than 2 Tx antennas cannot provide the rate of one (see Section 2.3.2.1 in this book or [Liang, 2003], [Liang and Xia, 2003]). However, they can still achieve the full diversity for the given number of Tx antennas. It is noted that we only consider the *square* CO STBCs in this chapter.

The *square* CO STBCs for eight Tx antennas proposed in the literature, which contain *numerous* zeros in the code matrices (see Eq. (2.36) in Chapter 2 of this book or [Tirkkonen and Hottinen, 2002] for instance), require a high peak-to-mean power ratio for the Tx antennas to achieve a certain bit error rate. This impedes their practical implementation.

In this chapter, three new square, order-8 CO STBCs with fewer zero entries compared to the conventional CO STBCs or even without zero entries are derived. These new complex-valued codes are more amenable to practical implementations as they allow for a more uniform spread of power among the

Tx antennas while requiring a lower peak power per Tx antenna to provide the same performance as other conventional, order-8 codes (with numerous zeros).

The content of this chapter is based on the published works [Seberry et al., 2004], [Tran et al., 2004a], [Zhao et al., 2005a]. We mark the discovery of the three new codes by two milestones in Fig. 1.1 of Chapter 1, which presents the history of CO STBCs.

The chapter is organized as follows. In Section 3.2, we introduce the new code designs. Section 3.3 provides expressions for Maximum Likelihood (ML) decoding of the transmitted symbols. Section 3.4 discusses the choice of signal constellation to provide optimal peak-to-mean power ratio. Section 3.5 presents some simulation results on the proposed CO STBCs, while Section 3.6 concludes the chapter.

3.2   New Complex Orthogonal Designs of Order Eight

The construction of CO STBCs follows directly from Complex Orthogonal Designs (CODs) which, in turn, can be constructed from AODs. The most general definitions of real orthogonal designs (ODs) and CODs (also CO STBCs) have been derived in Sections 2.3.1 and 2.3.2 respectively. However, since we consider here *square* CODs only, to facilitate our later discussions, we redefine the definitions of square, real ODs and *square* CODs (also *square* CO STBCs) as follows:

DEFINITION 3.2.1 *Let x_1, x_2,..., x_t be commuting, real indeterminates. A real orthogonal design $\mathbf{X}$ of order n and type $(a_1, a_2, \ldots, a_t)$, denoted as $OD(n; a_1, a_2, \ldots, a_t)$ where the coefficients a_i are positive integers, is a matrix of order n with entries chosen from $0, \pm x_1, \pm x_2, \ldots, \pm x_t$, such that*

$$\mathbf{X}^T\mathbf{X} \;=\; \sum_{i=1}^{t} a_i x_i^2 \mathbf{I}_n$$

where $\mathbf{X}^T$ denotes the transpose of the matrix $\mathbf{X}$ and $\mathbf{I}_n$ is the identity matrix of order n.

DEFINITION 3.2.2 *(Definition 5.4. in [Geramita and Seberry, 1979]) Let $\mathbf{X}$ be an $OD(n; u_1, \ldots, u_s)$ on the real variables $x_1, \ldots, x_s$ and let $\mathbf{Y}$ be an $OD(n; v_1, \ldots, v_t)$ on the real variables $y_1, \ldots, y_t$. It is said that $\mathbf{X}$ and $\mathbf{Y}$ are AODs $(n; u_1, \ldots, u_s; v_1, \ldots, v_t)$ if $\mathbf{X}\mathbf{Y}^T = \mathbf{Y}\mathbf{X}^T$.*

Table 3.1. Number of variables in an amicable pair with $n = 8$ [Geramita and Seberry, 1979]

Number of variables in **Y**	8	7	6	5	4	3	2	1	0
Number of variables in **X**	0	0	0	1	4	4	4	5	8

DEFINITION 3.2.3 *A square COD* $\mathbf{Z} = \mathbf{X} + i\mathbf{Y}$ *of order* n *is an* $n \times n$ *matrix on the complex indeterminates* $s_1, \cdots, s_t$, *with entries chosen from* $0, \pm s_1, \cdots, \pm s_t$, *their conjugates* $\pm s_1^*, \cdots, \pm s_t^*$, *or their products with* $i = \sqrt{-1}$, *such that*

$$\mathbf{Z}^H\mathbf{Z} = \Big(\sum_{k=1}^{t} |s_k|^2\Big)\mathbf{I}_n \tag{3.1}$$

where $\mathbf{Z}^H$ *denotes the Hermitian transpose of* $\mathbf{Z}$.

For the matrix $\mathbf{Z}$ to satisfy (3.1), the matrices $\mathbf{X}$ and $\mathbf{Y}$ must be a pair of AODs, which implies that both $\mathbf{X}$ and $\mathbf{Y}$ are orthogonal designs themselves and $\mathbf{X}\mathbf{Y}^T = \mathbf{Y}\mathbf{X}^T$.

It has been shown in [Geramita and Seberry, 1979] that, for order $n = 8$, the total number of different variables in the amicable pair $\mathbf{X}$ and $\mathbf{Y}$ cannot exceed eight. In Table 3.1, we record the number of variables in $\mathbf{X}$ versus the number of variables in $\mathbf{Y}$ with order eight.

It has been shown in [Street, 1982], that the construction of CODs can be facilitated by representing $\mathbf{Z}$ as

$$\mathbf{Z} = \sum_{j=1}^{t} \mathbf{A}_j s_j^R + i \sum_{j=1}^{t} \mathbf{B}_j s_j^I \tag{3.2}$$

where s_j^R and s_j^I denote the real and imaginary parts of the complex variables $s_j = s_j^R + i s_j^I$ and $\mathbf{A}_j$ and $\mathbf{B}_j$ are the real coefficient matrices for s_j^R and s_j^I, respectively. To satisfy (3.1), the matrices $\{\mathbf{A}_j\}$ and $\{\mathbf{B}_j\}$ of order n must satisfy the following conditions:

$$\begin{aligned}
\mathbf{A}_j\mathbf{A}_j^T &= \mathbf{I}, \ \mathbf{B}_j\mathbf{B}_j^T = \mathbf{I}, \ \forall j = 1, ..., t \\
\mathbf{A}_k\mathbf{A}_j^T &= -\mathbf{A}_j\mathbf{A}_k^T, \ \mathbf{B}_k\mathbf{B}_j^T = -\mathbf{B}_j\mathbf{B}_k^T, \ k \neq j \\
\mathbf{A}_k\mathbf{B}_j^T &= \mathbf{B}_j\mathbf{A}_k^T, \ \forall k, j = 1, ..., t
\end{aligned} \tag{3.3}$$

The conditions in (3.3) are necessary and sufficient for the existence of AODs of order n. Thus, the problem of finding CODs is connected to the theory of AODs.

From the perspective of constructing CO STBCs, the most promising case is that in which both $\mathbf{X}$ and $\mathbf{Y}$ have four variables. This case has been considered in the conventional, order-8 CO STBCs, corresponding to $COD(8; 1, 1, 1, 1)$ with all four variables appearing *once* in each column of $\mathbf{Z}$. An example is given in (3.4) (see [Tarokh et al., 1999b], [Tirkkonen and Hottinen, 2002], or Eq. (2.36) in Chapter 2):

$$
\mathbf{Z}_1 =
\begin{bmatrix}
s_1 & s_2 & s_3 & 0 & s_4 & 0 & 0 & 0 \\
-s_2^* & s_1^* & 0 & -s_3 & 0 & -s_4 & 0 & 0 \\
-s_3^* & 0 & s_1^* & s_2 & 0 & 0 & -s_4 & 0 \\
0 & s_3^* & -s_2^* & s_1 & 0 & 0 & 0 & s_4 \\
-s_4^* & 0 & 0 & 0 & s_1^* & s_2 & s_3 & 0 \\
0 & s_4^* & 0 & 0 & -s_2^* & s_1 & 0 & -s_3 \\
0 & 0 & s_4^* & 0 & -s_3^* & 0 & s_1 & s_2 \\
0 & 0 & 0 & -s_4^* & 0 & s_3^* & -s_2^* & s_1^*
\end{bmatrix}
\tag{3.4}
$$

These conventional codes contain numerous zero entries which are undesirable. Note that we use the similar notation to that mentioned in [Geramita and Seberry, 1979], i.e. $COD(8; 1, 1, 1, 1)$, to denote a square, order-8 COD containing four complex variables and each variable appearing once in each column. Readers may refer to [Geramita and Seberry, 1979] for more details.

In [Seberry et al., 2004], [Tran et al., 2004a], [Zhao et al., 2005a], we have introduced two new codes of order eight where some variables appear more often than others (more than once in each column), i.e., codes based on $COD(8; 1, 1, 2, 2)$ and $COD(8; 1, 1, 1, 4)$. These codes, namely $\mathbf{Z}_2$ and $\mathbf{Z}_3$, are given in (3.5) and (3.6), respectively. Details about the design methods for these two CODs are given below.

THEOREM 3.2.4 *(Proposition 5.1.1 in [Wolfe, 1975]) If there exist a pair of AODs $(n; a_1, \ldots, a_s; b_1, \ldots, b_t)$ and a pair of AODs $(m; c_1, \ldots, c_u; d_1, \ldots, d_v)$, then there exists a pair of AODs $(mn; b_1c_1, \ldots, b_1c_{u-1}, a_1c_u, \ldots, a_sc_u; b_1d_1, \ldots, b_1d_v, b_2c_u, \ldots, b_tc_u)$.*

Proof. Let $\mathbf{X} = \sum_{i=1}^{s} \mathbf{A}_i x_i$ and $\mathbf{Y} = \sum_{j=1}^{t} \mathbf{B}_j y_j$ be the amicable designs of order n and let $\mathbf{Z} = \sum_{k=1}^{u} \mathbf{C}_k z_k$ and $\mathbf{W} = \sum_{l=1}^{v} \mathbf{D}_l w_l$ be the amicable designs of order m. Construct the matrices

$$\mathbf{P} = \sum_{i=1}^{u-1}(\mathbf{B}_1 \bigotimes \mathbf{C}_i)p_i + \sum_{j=1}^{s}(\mathbf{A}_j \bigotimes \mathbf{C}_u)p_{j+u-1}$$

$$\mathbf{Q} = \sum_{i=1}^{v}(\mathbf{B}_1 \bigotimes \mathbf{D}_i)q_i + \sum_{j=2}^{t}(\mathbf{B}_j \bigotimes \mathbf{C}_u)q_{j+v-1}$$

where $\bigotimes$ is the Kronecker product, p_i for $i = 1, \ldots, s + u - 1$ and q_j for $j = 1, \ldots, t + v - 1$ are distinct commuting indeterminates. It can be proved that $\mathbf{P}$ and $\mathbf{Q}$ are a pair of AODs $(mn; b_1 c_1, \ldots, b_1 c_{u-1}, a_1 c_u, \ldots, a_s c_u; b_1 d_1, \ldots, b_1 d_v, b_2 c_u, \ldots, b_t c_u)$.

Using the above theorem, the AOD $(2; 1, 1; 1, 1)$ and the AOD $(4; 1, 1, 2; 1, 1, 2)$, we can construct the AOD $(8; 1, 1, 2, 2; 1, 1, 2, 2)$ which can be represented by the set of 8×8 coefficient matrices as follows (the signs "–" stand for "–1"):

$$\mathbf{P}_1 = \begin{bmatrix} 1 & 0 & 0 & 0 & 0 & 0 & 0 & 0 \\ 0 & 1 & 0 & 0 & 0 & 0 & 0 & 0 \\ 0 & 0 & - & 0 & 0 & 0 & 0 & 0 \\ 0 & 0 & 0 & - & 0 & 0 & 0 & 0 \\ 0 & 0 & 0 & 0 & 1 & 0 & 0 & 0 \\ 0 & 0 & 0 & 0 & 0 & 1 & 0 & 0 \\ 0 & 0 & 0 & 0 & 0 & 0 & - & 0 \\ 0 & 0 & 0 & 0 & 0 & 0 & 0 & - \end{bmatrix}, \quad \mathbf{Q}_1 = \begin{bmatrix} 1 & 0 & 0 & 0 & 0 & 0 & 0 & 0 \\ 0 & - & 0 & 0 & 0 & 0 & 0 & 0 \\ 0 & 0 & 0 & 1 & 0 & 0 & 0 & 0 \\ 0 & 0 & 1 & 0 & 0 & 0 & 0 & 0 \\ 0 & 0 & 0 & 0 & 1 & 0 & 0 & 0 \\ 0 & 0 & 0 & 0 & 0 & - & 0 & 0 \\ 0 & 0 & 0 & 0 & 0 & 0 & 0 & 1 \\ 0 & 0 & 0 & 0 & 0 & 0 & 1 & 0 \end{bmatrix}$$

$$\mathbf{P}_2 = \begin{bmatrix} 0 & 1 & 0 & 0 & 0 & 0 & 0 & 0 \\ - & 0 & 0 & 0 & 0 & 0 & 0 & 0 \\ 0 & 0 & 0 & - & 0 & 0 & 0 & 0 \\ 0 & 0 & 1 & 0 & 0 & 0 & 0 & 0 \\ 0 & 0 & 0 & 0 & 0 & 1 & 0 & 0 \\ 0 & 0 & 0 & 0 & - & 0 & 0 & 0 \\ 0 & 0 & 0 & 0 & 0 & 0 & 0 & - \\ 0 & 0 & 0 & 0 & 0 & 0 & 1 & 0 \end{bmatrix}, \quad \mathbf{Q}_2 = \begin{bmatrix} 0 & 1 & 0 & 0 & 0 & 0 & 0 & 0 \\ 1 & 0 & 0 & 0 & 0 & 0 & 0 & 0 \\ 0 & 0 & 1 & 0 & 0 & 0 & 0 & 0 \\ 0 & 0 & 0 & - & 0 & 0 & 0 & 0 \\ 0 & 0 & 0 & 0 & 0 & 1 & 0 & 0 \\ 0 & 0 & 0 & 0 & 1 & 0 & 0 & 0 \\ 0 & 0 & 0 & 0 & 0 & 0 & 1 & 0 \\ 0 & 0 & 0 & 0 & 0 & 0 & 0 & - \end{bmatrix}$$

$$\mathbf{P}_3 = \begin{bmatrix} 0 & 0 & 1 & 1 & 0 & 0 & 0 & 0 \\ 0 & 0 & 1 & - & 0 & 0 & 0 & 0 \\ 1 & 1 & 0 & 0 & 0 & 0 & 0 & 0 \\ 1 & - & 0 & 0 & 0 & 0 & 0 & 0 \\ 0 & 0 & 0 & 0 & 0 & 0 & - & - \\ 0 & 0 & 0 & 0 & 0 & 0 & - & 1 \\ 0 & 0 & 0 & 0 & - & - & 0 & 0 \\ 0 & 0 & 0 & 0 & - & 1 & 0 & 0 \end{bmatrix}, \quad \mathbf{Q}_3 = \begin{bmatrix} 0 & 0 & 1 & 1 & 0 & 0 & 0 & 0 \\ 0 & 0 & 1 & - & 0 & 0 & 0 & 0 \\ - & - & 0 & 0 & 0 & 0 & 0 & 0 \\ - & 1 & 0 & 0 & 0 & 0 & 0 & 0 \\ 0 & 0 & 0 & 0 & 0 & 0 & 1 & 1 \\ 0 & 0 & 0 & 0 & 0 & 0 & 1 & - \\ 0 & 0 & 0 & 0 & - & - & 0 & 0 \\ 0 & 0 & 0 & 0 & - & 1 & 0 & 0 \end{bmatrix}$$

$$
\mathbf{P}_4 = \begin{bmatrix}
0 & 0 & 0 & 0 & 0 & 0 & 1 & 1 \\
0 & 0 & 0 & 0 & 0 & 0 & 1 & - \\
0 & 0 & 0 & 0 & 1 & 1 & 0 & 0 \\
0 & 0 & 0 & 0 & 1 & - & 0 & 0 \\
0 & 0 & 1 & 1 & 0 & 0 & 0 & 0 \\
0 & 0 & 1 & - & 0 & 0 & 0 & 0 \\
1 & 1 & 0 & 0 & 0 & 0 & 0 & 0 \\
1 & - & 0 & 0 & 0 & 0 & 0 & 0
\end{bmatrix}, \quad
\mathbf{Q}_4 = \begin{bmatrix}
0 & 0 & 0 & 0 & 0 & 0 & 1 & 1 \\
0 & 0 & 0 & 0 & 0 & 0 & 1 & - \\
0 & 0 & 0 & 0 & 1 & 1 & 0 & 0 \\
0 & 0 & 0 & 0 & 1 & - & 0 & 0 \\
0 & 0 & - & - & 0 & 0 & 0 & 0 \\
0 & 0 & - & 1 & 0 & 0 & 0 & 0 \\
- & - & 0 & 0 & 0 & 0 & 0 & 0 \\
- & 1 & 0 & 0 & 0 & 0 & 0 & 0
\end{bmatrix}
$$

Let $\{s_j = s_j^R + is_j^I\}_{j=1}^4$ to be a set of complex symbols. By using the construction

$$
\mathbf{Z} = \sum_{j=1}^{4} \mathbf{P}_j s_j^R + i \sum_{j=1}^{4} \mathbf{Q}_j s_j^I
$$

we have the following orthogonal design:

$$
\mathbf{Z} = \begin{bmatrix}
s_1 & s_2 & s_3 & s_3 \\
-s_2^* & s_1^* & s_3 & -s_3 \\
s_3^* & s_3^* & -s_1^R + is_2^I & -s_2^R + is_1^I \\
s_3^* & -s_3^* & s_2^R + is_1^I & -s_1^R - is_2^I \\
0 & 0 & s_4^* & s_4^* \\
0 & 0 & s_4^* & -s_4^* \\
s_4^* & s_4^* & 0 & 0 \\
s_4^* & -s_4^* & 0 & 0
\end{bmatrix}
$$

$$
\begin{bmatrix}
0 & 0 & s_4 & s_4 \\
0 & 0 & s_4 & -s_4 \\
s_4 & s_4 & 0 & 0 \\
s_4 & -s_4 & 0 & 0 \\
s_1 & s_2 & -s_3^* & -s_3^* \\
-s_2^* & s_1^* & -s_3^* & s_3^* \\
-s_3^* & -s_3^* & -s_1^R + is_2^I & -s_2^R + is_1^I \\
-s_3^* & s_3^* & s_2^R + is_1^I & -s_1^R - is_2^I
\end{bmatrix}
$$

To achieve a COD following Definition 3.2.3, we can scale the symbols s_3 and s_4 by a factor of $1/\sqrt{2}$. We denote the resultant COD(8;1,1,2,2) to be $\mathbf{Z}_2$ which is given in (3.5).

DEFINITION 3.2.5 (*Definition 2.1. in [Geramita and Seberry, 1979]) A weighting matrix* $\mathbf{W}(n, k)$ *of weight* k *and order* n *is an* $n \times n$ *matrix* $\mathbf{W}$ *with the entries chosen from* $\{0, \pm 1\}$ *such that*

$$
\mathbf{W}^T \mathbf{W} = k \mathbf{I}_n
$$

If $n = k$, then $\mathbf{W}(n, k)$ is usually referred to as an Hadamard matrix in the literature.

$$
\mathbf{Z}_2 =
\begin{bmatrix}
s_1 & s_2 & \frac{s_3}{\sqrt{2}} & \frac{s_3}{\sqrt{2}} \\
-s_2^* & s_1^* & \frac{s_3}{\sqrt{2}} & -\frac{s_3}{\sqrt{2}} \\
\frac{s_3^*}{\sqrt{2}} & \frac{s_3^*}{\sqrt{2}} & -s_1^R + is_2^I & -s_2^R + is_1^I \\
\frac{s_3^*}{\sqrt{2}} & -\frac{s_3^*}{\sqrt{2}} & s_2^R + is_1^I & -s_1^R - is_2^I \\
0 & 0 & \frac{s_4^*}{\sqrt{2}} & \frac{s_4^*}{\sqrt{2}} \\
0 & 0 & \frac{s_4^*}{\sqrt{2}} & -\frac{s_4^*}{\sqrt{2}} \\
\frac{s_4^*}{\sqrt{2}} & \frac{s_4^*}{\sqrt{2}} & 0 & 0 \\
\frac{s_4^*}{\sqrt{2}} & -\frac{s_4^*}{\sqrt{2}} & 0 & 0 \\
\end{bmatrix}
$$

$$
\begin{matrix}
0 & 0 & \frac{s_4}{\sqrt{2}} & \frac{s_4}{\sqrt{2}} \\
0 & 0 & \frac{s_4}{\sqrt{2}} & -\frac{s_4}{\sqrt{2}} \\
\frac{s_4}{\sqrt{2}} & \frac{s_4}{\sqrt{2}} & 0 & 0 \\
\frac{s_4}{\sqrt{2}} & -\frac{s_4}{\sqrt{2}} & 0 & 0 \\
s_1 & s_2 & -\frac{s_3^*}{\sqrt{2}} & -\frac{s_3^*}{\sqrt{2}} \\
-s_2^* & s_1^* & -\frac{s_3^*}{\sqrt{2}} & \frac{s_3^*}{\sqrt{2}} \\
-\frac{s_3^*}{\sqrt{2}} & -\frac{s_3^*}{\sqrt{2}} & -s_1^R + is_2^I & -s_2^R + is_1^I \\
-\frac{s_3^*}{\sqrt{2}} & \frac{s_3^*}{\sqrt{2}} & s_2^R + is_1^I & -s_1^R - is_2^I \\
\end{matrix}
\qquad (3.5)
$$

THEOREM 3.2.6 *([Street, 1981]) Suppose that $(\mathbf{A}, \mathbf{B})$ and $(\mathbf{C}, \mathbf{D})$ are both AODs $(n; a_1, \ldots, a_s; b_1, \ldots, b_t)$. Suppose further that there exits a weighting matrix $\mathbf{W}(n, k)$ such that*

$$
\mathbf{A}\mathbf{W}^T = \mathbf{W}\mathbf{C}^T, \quad \mathbf{B}\mathbf{W}^T = -\mathbf{W}\mathbf{D}^T
$$

Then there exits AODs $(2n; k, a_1, \ldots, a_s; k, b_1, \ldots, b_t)$.

Using the above theorem, with the given $\mathbf{W}(4, 4)$ and AODs $(4; 1, 1, 1; 1, 1, 1)$ as follows:

$$
\mathbf{W} =
\begin{bmatrix}
1 & 1 & 1 & 1 \\
1 & -1 & 1 & -1 \\
1 & 1 & -1 & -1 \\
1 & -1 & -1 & 1 \\
\end{bmatrix},
$$

$$
\mathbf{A} = \begin{bmatrix} x_1 & 0 & x_3 & x_2 \\ 0 & x_1 & -x_2 & x_3 \\ -x_3 & x_2 & x_1 & 0 \\ -x_2 & -x_3 & 0 & x_1 \end{bmatrix}, \quad
\mathbf{B} = \begin{bmatrix} y_1 & 0 & y_2 & y_3 \\ 0 & y_1 & y_3 & -y_2 \\ y_2 & y_3 & -y_1 & 0 \\ y_3 & -y_2 & 0 & -y_1 \end{bmatrix},
$$

$$
\mathbf{C} = \begin{bmatrix} x_1 & x_2 & x_3 & 0 \\ -x_2 & x_1 & 0 & x_3 \\ -x_3 & 0 & x_1 & -x_2 \\ 0 & -x_3 & x_2 & x_1 \end{bmatrix}, \quad
\mathbf{D} = \begin{bmatrix} -y_3 & -y_2 & -y_1 & 0 \\ -y_2 & y_3 & 0 & -y_1 \\ -y_1 & 0 & y_3 & y_2 \\ 0 & -y_1 & y_2 & -y_3 \end{bmatrix}
$$

we construct the following AODs $(8;1,1,1,4;1,1,1,4)$ for order 8:

$$
\mathbf{X} = \begin{bmatrix} \mathbf{A} & \mathbf{W}x_4 \\ -\mathbf{W}x_4 & \mathbf{C} \end{bmatrix}
$$

$$
= \begin{bmatrix}
x_1 & 0 & x_3 & x_2 & x_4 & x_4 & x_4 & x_4 \\
0 & x_1 & -x_2 & x_3 & x_4 & -x_4 & x_4 & -x_4 \\
-x_3 & x_2 & x_1 & 0 & x_4 & x_4 & -x_4 & -x_4 \\
-x_2 & -x_3 & 0 & x_1 & x_4 & -x_4 & -x_4 & x_4 \\
-x_4 & -x_4 & -x_4 & -x_4 & x_1 & x_2 & x_3 & 0 \\
-x_4 & x_4 & -x_4 & x_4 & -x_2 & x_1 & 0 & x_3 \\
-x_4 & -x_4 & x_4 & x_4 & -x_3 & 0 & x_1 & -x_2 \\
-x_4 & x_4 & x_4 & -x_4 & 0 & -x_3 & x_2 & x_1
\end{bmatrix},
$$

$$
\mathbf{Y} = \begin{bmatrix} \mathbf{B} & \mathbf{W}y_4 \\ \mathbf{W}y_4 & \mathbf{D} \end{bmatrix}
$$

$$
= \begin{bmatrix}
y_1 & 0 & y_2 & y_3 & y_4 & y_4 & y_4 & y_4 \\
0 & y_1 & y_3 & -y_2 & y_4 & -y_4 & y_4 & -y_4 \\
y_2 & y_3 & -y_1 & 0 & y_4 & y_4 & -y_4 & -y_4 \\
0 & -x_3 & x_2 & x_1 & y_4 & -y_4 & -y_4 & y_4 \\
y_4 & y_4 & y_4 & y_4 & -y_3 & -y_2 & -y_1 & 0 \\
y_4 & -y_4 & y_4 & -y_4 & -y_2 & y_3 & 0 & -y_1 \\
y_4 & y_4 & -y_4 & -y_4 & -y_1 & 0 & y_3 & y_2 \\
y_4 & -y_4 & -y_4 & y_4 & 0 & -y_1 & y_2 & -y_3
\end{bmatrix}
$$

Let $\mathbf{A_j}$ and $\mathbf{B_j}$ for $j = 1, \ldots, 4$ be the coefficient matrices of x_j and y_j, and denote $s_j = s_j^R + i s_j^I$ to be a set of complex symbols. By using the construction

$$
\mathbf{Z} = \sum_{j=1}^{4} \mathbf{A}_j s_j^R + i \sum_{j=1}^{4} \mathbf{B}_j s_j^I
$$

we have the following orthogonal design:

$$
\mathbf{Z} = \left[
\begin{array}{cccc}
s_1 & 0 & s_3^R + is_2^I & s_2^R + is_3^I \\
0 & s_1 & -s_2^R + is_3^I & s_3^R - is_2^I \\
-s_3^R + is_2^I & s_2^R + is_3^I & s_1^* & 0 \\
-s_2^R + is_3^I & -s_3^R - is_2^I & 0 & s_1^* \\
-s_4^* & -s_4^* & -s_4^* & -s_4^* \\
-s_4^* & s_4^* & -s_4^* & s_4^* \\
-s_4^* & -s_4^* & s_4^* & s_4^* \\
-s_4^* & s_4^* & s_4^* & -s_4^* \\
\end{array}
\right.
$$

$$
\left.
\begin{array}{cccc}
s_4 & s_4 & s_4 & s_4 \\
s_4 & -s_4 & s_4 & -s_4 \\
s_4 & s_4 & -s_4 & -s_4 \\
s_4 & s_4 & -s_4 & s_4 \\
s_1^R - is_3^I & s_2^* & s_3^R - is_1^I & 0 \\
-s_2 & s_1^R + is_3^I & 0 & s_3^R - is_1^I \\
-s_3^R - is_1^I & 0 & s_1^R + is_3^I & -s_2^* \\
0 & -s_3^R - is_1^I & s_2 & s_1^R - is_3^I \\
\end{array}
\right]
$$

To have a COD following Definition 3.2.3, we scale the symbols s_4 by a factor of $1/2$. We denote the resultant COD(8;1,1,1,4) to be $\mathbf{Z}_3$ which is given in (3.6).

Various AODs cannot be constructed by general known theorems. A complete searching method for AODs existing for order 8 has not been derived yet. In [Street, 1982], it has been envisaged that a $COD(8; 2, 2, 2, 2)$ exists. However, no construction of it has been derived in the literature. Therefore, in [Seberry et al., 2004], [Zhao et al., 2005a], the code of order eight based on the $COD(8; 2, 2, 2, 2)$ has been proposed as given in (3.7). This code is in turn constructed based on the AOD $(8; 2, 2, 2, 2; 2, 2, 2, 2)$ by using an exhaustive search for the equivalent classes of orthogonal designs. The detailed description about this searching method can be found in [Zhao et al., 2005b].

All of the CO STBCs proposed here achieve the *maximum* code rate for order-8, *square* CO STBCs, which is equal to 1/2. We would like to recall that, according to Liang's paper [Liang, 2003], the maximum achievable rate of CO STBCs for $n = 2m - 1$ or $n = 2m$ Tx antennas is $R_{max} = (m + 1)/2m$. Particularly, for $n = 8$, i.e., $m = 4$, the maximum achievable rate of CO STBCs is 5/8.

$$
\mathbf{Z}_3 =
\begin{bmatrix}
s_1 & 0 & s_3^R+is_2^I & s_2^R+is_3^I & \frac{s_4}{2} & \frac{s_4}{2} & \frac{s_4}{2} & \frac{s_4}{2} \\[4pt]
0 & s_1 & -s_2^R+is_3^I & s_3^R-is_2^I & \frac{s_4}{2} & -\frac{s_4}{2} & \frac{s_4}{2} & -\frac{s_4}{2} \\[4pt]
-s_3^R+is_2^I & s_2^R+is_3^I & s_1^* & 0 & \frac{s_4}{2} & \frac{s_4}{2} & -\frac{s_4}{2} & -\frac{s_4}{2} \\[4pt]
-s_2^R+is_3^I & -s_3^R-is_2^I & 0 & s_1^* & \frac{s_4}{2} & \frac{s_4}{2} & -\frac{s_4}{2} & \frac{s_4}{2} \\[4pt]
-\frac{s_4^*}{2} & -\frac{s_4^*}{2} & -\frac{s_4^*}{2} & -\frac{s_4^*}{2} & s_1^R-is_3^I & s_2^* & s_3^R-is_1^I & 0 \\[4pt]
-\frac{s_4^*}{2} & \frac{s_4^*}{2} & -\frac{s_4^*}{2} & \frac{s_4^*}{2} & -s_2 & s_1^R+is_3^I & 0 & s_3^R-is_1^I \\[4pt]
-\frac{s_4^*}{2} & -\frac{s_4^*}{2} & \frac{s_4^*}{2} & \frac{s_4^*}{2} & -s_3^R-is_1^I & 0 & s_1^R+is_3^I & -s_2^* \\[4pt]
-\frac{s_4^*}{2} & \frac{s_4^*}{2} & \frac{s_4^*}{2} & -\frac{s_4^*}{2} & 0 & -s_3^R-is_1^I & s_2 & s_1^R-is_3^I
\end{bmatrix}
\tag{3.6}
$$

However, this maximum rate is only achievable for *non-square* constructions. For *square* constructions of orders $n = 2^a(2b+1)$, the maximum achievable rate is $R_{max} = (a+1)/\left[2^a(2b+1)\right]$. For $n = 8$, i.e., $a = 3$ and $b = 0$, the maximum achievable rate of *square* CO STBCs is only 1/2.

The *vague* statement on the maximum achievable rate of CO STBCs in Liang's paper [Liang, 2003], which easily makes readers confused, has been pointed out in Chapter 2 of the book. Readers are recommended to refer to Remark 2.3.8 in Chapter 2 for more details.

A question that could be raised is why *square* CO STBCs are of particular interest in this chapter. It is because, *square* CO STBCs have a great advantage over *non-square* CO STBCs in that they require a much smaller length of the codes, i.e., much smaller processing delay, though, the maximum rate of the former may be smaller than that of the later.

$$\mathbf{Z}_4 = \left[\begin{array}{cccccccc}
\frac{s_1}{\sqrt{2}} & \frac{s_1}{\sqrt{2}} & \frac{s_2}{\sqrt{2}} & \frac{s_2}{\sqrt{2}} & \frac{s_3}{\sqrt{2}} & \frac{s_4}{\sqrt{2}} & \frac{s_3}{\sqrt{2}} & \frac{s_4}{\sqrt{2}} \\[2mm]
\frac{s_1}{\sqrt{2}} & -\frac{s_1}{\sqrt{2}} & \frac{s_2}{\sqrt{2}} & -\frac{s_2}{\sqrt{2}} & \frac{s_4^*}{\sqrt{2}} & -\frac{s_3^*}{\sqrt{2}} & \frac{s_4^*}{\sqrt{2}} & -\frac{s_3^*}{\sqrt{2}} \\[2mm]
\frac{s_2^*}{\sqrt{2}} & \frac{s_2^*}{\sqrt{2}} & -\frac{s_1^*}{\sqrt{2}} & -\frac{s_1^*}{\sqrt{2}} & \frac{s_3}{\sqrt{2}} & \frac{s_4}{\sqrt{2}} & -\frac{s_3}{\sqrt{2}} & -\frac{s_4}{\sqrt{2}} \\[2mm]
\frac{s_2^*}{\sqrt{2}} & -\frac{s_2^*}{\sqrt{2}} & -\frac{s_1^*}{\sqrt{2}} & \frac{s_1^*}{\sqrt{2}} & \frac{s_4^*}{\sqrt{2}} & -\frac{s_3^*}{\sqrt{2}} & -\frac{s_4^*}{\sqrt{2}} & \frac{s_3^*}{\sqrt{2}} \\[2mm]
\frac{-s_4^R+is_3^I}{\sqrt{2}} & \frac{-s_3^R+is_4^I}{\sqrt{2}} & \frac{-s_4^R+is_3^I}{\sqrt{2}} & \frac{-s_3^R+is_4^I}{\sqrt{2}} & \frac{s_2^R-is_1^I}{\sqrt{2}} & \frac{s_2^R-is_1^I}{\sqrt{2}} & \frac{s_1^R-is_2^I}{\sqrt{2}} & \frac{s_1^R-is_2^I}{\sqrt{2}} \\[2mm]
\frac{-s_3^R-is_4^I}{\sqrt{2}} & \frac{s_4^R+is_3^I}{\sqrt{2}} & \frac{-s_3^R-is_4^I}{\sqrt{2}} & \frac{s_4^R+is_3^I}{\sqrt{2}} & \frac{s_2^R-is_1^I}{\sqrt{2}} & \frac{-s_2^R+is_1^I}{\sqrt{2}} & \frac{s_1^R-is_2^I}{\sqrt{2}} & \frac{-s_1^R+is_2^I}{\sqrt{2}} \\[2mm]
\frac{-s_4^R+is_3^I}{\sqrt{2}} & \frac{-s_3^R+is_4^I}{\sqrt{2}} & \frac{s_4^R-is_3^I}{\sqrt{2}} & \frac{s_3^R-is_4^I}{\sqrt{2}} & \frac{s_1^R+is_2^I}{\sqrt{2}} & \frac{s_1^R+is_2^I}{\sqrt{2}} & \frac{-s_2^R-is_1^I}{\sqrt{2}} & \frac{-s_2^R-is_1^I}{\sqrt{2}} \\[2mm]
\frac{-s_3^R-is_4^I}{\sqrt{2}} & \frac{s_4^R+is_3^I}{\sqrt{2}} & \frac{s_3^R+is_4^I}{\sqrt{2}} & \frac{-s_4^R-is_3^I}{\sqrt{2}} & \frac{s_1^R+is_2^I}{\sqrt{2}} & \frac{-s_1^R-is_2^I}{\sqrt{2}} & \frac{-s_2^R-is_1^I}{\sqrt{2}} & \frac{s_2^R+is_1^I}{\sqrt{2}}
\end{array}\right] \quad (3.7)$$

Let us consider CO STBCs for $n = 8$ Tx antennas as an example (also see Example 2.3.1 in Chapter 2). The *non-square* CO STBC that achieves the maximum rate 5/8 requires the length of 112 symbol time slots (STSs) as shown by Table 2.6 in Chapter 2 of this book. The [112,8,70] CO STBC given in Appendix E in Liang's paper [Liang, 2003] is an example for this case. As opposed to *non-square* CO STBCs, *square* CO STBCs only require the length of 8 STSs to achieve the maximum rate 1/2, which is slightly smaller than the maximum rate of *non-square* CO STBCs. Clearly, *square* CO STBCs require a much shorter length, especially for a large number of Tx antennas, with the consequence of a slightly lower code rate. For this reason, *square* CO STBCs are of our particular interest.

Apart from having the maximum rate, our proposed CO STBCs $\mathbf{Z}_2$, $\mathbf{Z}_3$, $\mathbf{Z}_4$ have fewer zero entries (compared to the conventional codes) or even no zero entries in the code matrices. This property results in a more uniform transmission power distribution between Tx antennas. Intuitively, due to this property, our proposed CO STBCs require a lower peak power per Tx antenna to achieve the same bit error performance as the conventional CO STBCs containing numerous zeros. Equivalently, with the same peak power at Tx antennas, our proposed codes provide a better bit error performance than the conventional CO STBCs.

In addition, our codes are more amenable to practical implementation than the conventional code, since, transmitter antennas are turned off less frequently or even are not required to be turned off during transmission unlike with the conventional codes.

3.3   Decoding Metrics

In this section, a channel comprising 8 Tx antennas and 1 Rx antenna is examined. Let $\mathbf{R}_{1\times 8}$, $\mathbf{H}_{1\times 8}$ and $\mathbf{N}_{1\times 8}$ be the matrices of received signals, of channel coefficients and of noise, respectively. The Tx antennas are assumed to be sufficiently separate, so that the channel coefficients between these Tx antennas and the Rx antenna are independent of one another. The transmission model is then as follows:

$$\mathbf{R} = \mathbf{HZ} + \mathbf{N}$$

The transmitted symbols can be decoded following the Maximum Likelihood (ML) decoding scheme, which is expressed as

$$\{\hat{s}_k\}_{k=1}^4 = Arg\left(\min_{\{s_k\},s_k \in S} \|\mathbf{R} - \mathbf{HZ}\|_F\right) \tag{3.8}$$

where $Arg(x)$ is the argument of x; $\|\Re\|_F$ is the Frobenius norm of the matrix $\Re$, i.e., the square root of the sum of all the magnitude squared elements of the matrix, and S is the set of all possible transmitted symbols.

We consider the conventional code $\mathbf{Z}_1$ first. Since the transmitted block code $\mathbf{Z}_1$ is orthogonal, (3.8) can be converted to four independent expressions for

Table 3.2. Decision metrics for decoding code $\mathbf{Z}_1$.

Variable	Decoding metric
s_1	$Arg\left\{ \min_{s_1 \in S} \left[\left\|(h_5 r_5^* + h_2 r_2^* + h_1^* r_1 + h_4^* r_4 + h_3 r_3^* + h_8 r_8^* \right.\right.\right.$ $\left.\left.\left. +h_6^* r_6 + h_7^* r_7) - s_1 \right\|^2 + \left(-1 + \sum_{i=1}^{8} |h_i|^2\right)|s_1|^2 \right] \right\}$
s_2	$Arg\left\{ \min_{s_2 \in S} \left[\left\|(-h_8 r_7^* - h_2 r_1^* - h_6 r_5^* + h_5^* r_6 + h_7^* r_8 + h_1^* r_2 \right.\right.\right.$ $\left.\left.\left. +h_3^* r_4 - h_4 r_3^*) - s_2 \right\|^2 + \left(-1 + \sum_{i=1}^{8} |h_i|^2\right)|s_2|^2 \right] \right\}$
s_3	$Arg\left\{ \min_{s_3 \subset S} \left[\left\|(h_8 r_6^* - h_3 r_1^* + h_1^* r_3 - h_2^* r_4 + h_4 r_2^* - h_7 r_5^* \right.\right.\right.$ $\left.\left.\left. +h_5^* r_7 - h_6^* r_8) - s_3 \right\|^2 + \left(-1 + \sum_{i=1}^{8} |h_i|^2\right)|s_3|^2 \right] \right\}$
s_4	$Arg\left\{ \min_{s_4 \in S} \left[\left\|(-h_8 r_4^* + h_7 r_3^* - h_5 r_1^* + h_6 r_2^* + h_1^* r_5 - h_2^* r_6 \right.\right.\right.$ $\left.\left.\left. -h_3^* r_7 + h_4^* r_8) - s_4 \right\|^2 + \left(-1 + \sum_{i=1}^{8} |h_i|^2\right)|s_4|^2 \right] \right\}$

four corresponding symbols. The decision metrics for decoding the symbols of code $\mathbf{Z}_1$ are given in Table 3.2. Similarly, the decision metrics for decoding codes $\mathbf{Z}_2$, $\mathbf{Z}_3$ and $\mathbf{Z}_4$ are derived in Tables 3.3, 3.4 and 3.5, respectively.

3.4 Choice of Signal Constellations

By examining the constituent matrices $\{\mathbf{A}_j\}_{j=1}^4$, $\{\mathbf{B}_j\}_{j=1}^4$, and the encoding matrix $\mathbf{Z}_4$, for instance, it is easy to notice that the entries z_{lk}, for $l = 5, \ldots, 8$ and $k = 1, \ldots, 8$, of $\mathbf{Z}_4$ are composed of the real part of one indeterminate and the imaginary part of another indeterminate, e.g., $z_{51} = \frac{-s_4^R + i s_3^I}{\sqrt{2}}$. This observation means that if the indeterminates $s_1, \ldots, s_4$ are chosen from the complex signal constellations where s_j^R or s_j^I ($j = 1, \ldots, 4$) can be equal to zero, e.g., the QPSK constellation $(1, -1, i, -i)$ then, some of the entries of the matrix $\mathbf{Z}_4$ can be equal to zero depending on the transmitted data. Therefore, such constellations should be avoided. An example of the constellation where the power is evenly spread among the Tx antennas independently of the transmitted data is the QPSK constellation $(1 + i, 1 - i, -1 + i, -1 - i)$. This observation also holds for the other proposed codes $\mathbf{Z}_2$ and $\mathbf{Z}_3$.

Table 3.3. Decision metrics for decoding code $\mathbf{Z}_2$.

Variable	Decoding metric
s_1	$Arg\left\{ \min_{s_1 \in S} \left[\left\lvert \frac{1}{2}\left(-h_4 r_4^* - h_3 r_4^* - h_4^* r_4 + h_3^* r_4 - h_8^* r_8 \right.\right.\right.\right.$ $-h_8 r_7^* - h_7 r_7^* + h_8^* r_7 - h_7^* r_7 - h_3 r_3^* - h_7 r_8^* - h_8 r_8^* + h_7^* r_8$ $\left.+2h_5^* r_5 + 2h_6 r_6^* + 2h_2 r_2^* + 2h_1^* r_1 - h_3^* r_3 + h_4^* r_3 - h_4 r_3^* \right) - s_1 \Bigg\rvert^2$ $\left.\left.+\left(-1 + \sum_{i=1}^{8} \lvert h_i \rvert^2 \right)\lvert s_1 \rvert^2 \right] \right\}$
s_2	$Arg\left\{ \min_{s_2 \in S} \left[\left\lvert \frac{1}{2}\left(h_4 r_4^* - h_3 r_4^* - h_4^* r_4 - h_3^* r_4 - 2h_6 r_5^* \right.\right.\right.\right.$ $+2h_5^* r_6 - h_7 r_7^* - h_7 r_8^* - h_7^* r_8 + h_8 r_8^* - 2h_2 r_1^* + h_8 r_7^* - h_3 r_3^*$ $\left.+2h_1^* r_2 + h_3^* r_3 + h_4^* r_3 + h_4 r_3^* - h_8^* r_8 + h_7^* r_7 + h_8^* r_7 \right) - s_2 \Bigg\rvert^2$ $\left.\left.+\left(-1 + \sum_{i=1}^{8} \lvert h_i \rvert^2 \right)\lvert s_2 \rvert^2 \right] \right\}$
s_3	$Arg\left\{ \min_{s_3 \in S} \left[\left\lvert \frac{1}{\sqrt{2}}\left(h_1^* r_4 - h_2^* r_4 + h_8^* r_6 - h_7^* r_5 + h_1^* r_3 \right.\right.\right.\right.$ $\left.-h_7^* r_6 - h_5 r_7^* - h_6 r_7^* - h_5 r_8^* + h_6 r_8^* + h_3 r_1^* + h_4 r_1^* + h_3 r_2^* - h_4 r_2^* \right.$ $\left.\left.\left.-h_8^* r_5 + h_2^* r_3 \right) - s_3 \Bigg\rvert^2 + \left(-1 + \sum_{i=1}^{8} \lvert h_i \rvert^2 \right)\lvert s_3 \rvert^2 \right] \right\}$
s_4	$Arg\left\{ \min_{s_4 \in S} \left[\left\lvert \frac{1}{\sqrt{2}}\left(-h_4^* r_6 + h_2^* r_7 + h_1^* r_7 - h_2^* r_8 + h_5 r_4^* \right.\right.\right.\right.$ $\left.+h_4^* r_5 + h_1^* r_8 + h_7 r_1^* + h_8 r_1^* + h_7 r_2^* - h_6 r_4^* + h_3^* r_5 + h_3^* r_6 - h_8 r_2^* \right.$ $\left.\left.\left.+h_5 r_3^* + h_6 r_3^* \right) - s_4 \Bigg\rvert^2 + \left(-1 + \sum_{i=1}^{8} \lvert h_i \rvert^2 \right)\lvert s_4 \rvert^2 \right] \right\}$

Table 3.4. Decision metrics for decoding code $\mathbf{Z}_3$.

Variable	Decoding metric
s_1	$Arg\left\{ \min_{s_1 \in S} \left[\left\| \frac{1}{2}\left(h_5 r_5^* + h_5^* r_5 - h_7^* r_5 + 2h_2^* r_2 + 2h_1^* r_1 \right. \right. \right. \right.$ $+2h_4 r_4^* + 2h_3 r_3^* + h_5 r_7^* + h_6 r_6^* + h_8 r_6^* - h_8^* r_6 + h_6^* r_6 + h_7 r_5^*$ $\left. +h_6 r_8^* + h_8 r_8^* - h_6^* r_8 + h_8^* r_8 + h_7 r_7^* - h_5^* r_7 + h_7^* r_7 \right) - s_1 \Big|^2$ $\left. \left. +\left(-1 + \sum_{i=1}^8 |h_i|^2 \right)|s_1|^2 \right] \right\}$
s_2	$Arg\left\{ \min_{s_2 \in S} \left[\left\| \frac{1}{2}\left(-2h_6^* r_5 + h_2 r_4^* + h_1 r_4^* - h_2^* r_4 + h_1 r_4^* \right. \right. \right. \right.$ $+h_3^* r_2 - h_4^* r_2 - h_3 r_1^* - h_4 r_1^* + h_3^* r_1 - h_4^* r_1 + h_1^* r_3 - h_2^* r_3$ $\left. -h_1 r_3^* - h_2 r_3^* + 2h_5 r_6^* + h_3 r_2^* + h_4 r_2^* - 2h_7 r_8^* + 2h_8^* r_7 \right) - s_2 \Big|^2$ $\left. \left. +\left(-1 + \sum_{i=1}^8 |h_i|^2 \right)|s_2|^2 \right] \right\}$
s_3	$Arg\left\{ \min_{s_3 \in S} \left[\left\| \frac{1}{2}\left(h_1^* r_4 + h_2^* r_4 - h_7 r_5^* - h_5^* r_5 - h_7^* r_5 \right. \right. \right. \right.$ $-h_1 r_4^* + h_2 r_4^* + h_3^* r_2 - h_4^* r_2 - h_3 r_1^* - h_4 r_1^* + h_4^* r_1 - h_3^* r_1$ $+h_1^* r_3 + h_2^* r_3 + h_5 r_7^* + h_1 r_3^* - h_2 r_3^* + h_5 r_5^* - h_3 r_2^* - h_4 r_2^*$ $-h_6 r_6^* - h_8 r_6^* + h_6^* r_6 - h_8^* r_6 + h_8 r_8^* + h_6^* r_8 - h_8^* r_8 - h_7 r_7^*$ $+h_6 r_8^* + h_5^* r_7 + h_7^* r_7 \Big) - s_3 \Big|^2 + \left(-1 + \sum_{i=1}^8 |h_i|^2 \right)|s_3|^2 \Big] \Big\}$
s_4	$Arg\left\{ \min_{s_4 \in S} \left[\left\| \frac{1}{2}\left(h_7 r_4^* + h_1^* r_5 + h_2^* r_5 + h_3^* r_5 + h_4^* r_5 + h_6 r_4^* \right. \right. \right. \right.$ $-h_8 r_4^* - h_5 r_3^* - h_5 r_1^* - h_6 r_1^* - h_7 r_1^* - h_8 r_1^* - h_6 r_3^* + h_8 r_3^*$ $+h_6 r_2^* - h_7 r_2^* + h_8 r_2^* - h_5 r_2^* - h_5 r_4^* + h_1^* r_6 - h_2^* r_6 + h_3^* r_6$ $-h_4^* r_6 - h_2^* r_8 - h_3^* r_8 + h_4^* r_8 + h_1^* r_7 + h_2^* r_7 - h_3^* r_7 - h_4^* r_7$ $+h_1^* r_8 + h_7 r_3^* \Big) - s_4 \Big|^2 + \left(-1 + \sum_{i=1}^8 |h_i|^2 \right)|s_4|^2 \Big] \Big\}$

Table 3.5.   Decision metrics for decoding code $\mathbf{Z}_4$.

Variable	Decoding metric
s_1	$Arg\left\{\min_{s_1 \in S}\left[\left\|\frac{1}{2\sqrt{2}}\left(-2h_3 r_4^* + 2h_4 r_4^* + h_5 r_5^* + h_6 r_5^*\right.\right.\right.\right.$ $+h_7 r_5^* + h_8 r_5^* + h_5 r_6^* - h_6 r_6^* + h_7 r_6^* + h_6 r_7^* + h_7 r_7^* + h_8 r_7^*$ $+h_5 r_7^* - h_8 r_6^* + h_7 r_8^* - h_8 r_8^* + h_5 r_8^* - h_6 r_8^* - 2h_3 r_3^* - 2h_4 r_3^*$ $+2h_1^* r_2 - 2h_2^* r_2 - h_5^* r_5 - h_6^* r_5 + h_7^* r_5 + h_8^* r_5 - h_5^* r_6 + h_6^* r_6$ $+h_7^* r_6 - h_8^* r_6 + h_5^* r_7 + h_6^* r_7 - h_7^* r_7 - h_8^* r_7 + h_5^* r_8 - h_6^* r_8$ $\left.-h_7^* r_8 + h_8^* r_8 + 2h_1^* r_1 + 2h_2^* r_1\right) - s_1\left.\right\|^2 + \left(-1 + \sum_{i=1}^{8}|h_i|^2\right)|s_1|^2\left.\right]\right\}$
s_2	$Arg\left\{\min_{s_2 \in S}\left[\left\|\frac{1}{2\sqrt{2}}\left(2h_4 r_1^* - 2h_4 r_2^* + 2h_3 r_2^* + h_6 r_5^*\right.\right.\right.\right.$ $-h_7 r_5^* + h_5 r_5^* + h_5 r_6^* - h_6 r_6^* - h_7 r_6^* - h_8 r_5^* + h_6 r_7^* - h_7 r_7^*$ $-h_8 r_7^* + h_5 r_7^* + h_8 r_6^* - h_7 r_8^* + h_8 r_8^* - h_6 r_8^* + h_5 r_8^* + 2h_3 r_1^*$ $+2h_1^* r_4 - 2h_2^* r_4 + h_5^* r_5 + h_6^* r_5 + h_7^* r_5 + h_8^* r_5 + h_5^* r_6 - h_6^* r_6$ $+h_7^* r_6 - h_8^* r_6 - h_5^* r_7 - h_6^* r_7 - h_7^* r_7 - h_8^* r_7 - h_5^* r_8 + h_6^* r_8$ $\left.-h_7^* r_8 + h_8^* r_8 + 2h_1^* r_3 + 2h_2^* r_3\right) - s_2\left.\right\|^2 + \left(-1 + \sum_{i=1}^{8}|h_i|^2\right)|s_2|^2\left.\right]\right\}$
s_3	$Arg\left\{\min_{s_3 \in S}\left[\left\|\frac{1}{2\sqrt{2}}\left(-2h_4 r_6^* + h_7 r_3^* + h_8 r_3^* - h_5 r_1^* - h_5 r_4^*\right.\right.\right.\right.$ $-h_6 r_1^* - 2h_2 r_6^* - h_7 r_1^* - h_6 r_4^* + h_7 r_4^* - h_5 r_2^* + h_8 r_4^* - h_7 r_2^* - h_6 r_2^*$ $-h_8 r_2^* - h_8 r_1^* + 2h_4 r_8^* - 2h_2 r_8^* - h_5 r_3^* - h_6 r_3^* - h_5^* r_2 - h_5^* r_4$ $+h_6^* r_4 + h_7^* r_4 - h_8^* r_4 + h_6^* r_2 - h_7^* r_2 + h_8^* r_2 + 2h_1^* r_5 + 2h_3^* r_5$ $+2h_1^* r_7 - 2h_3^* r_7 + h_5^* r_3 - h_6^* r_3 - h_7^* r_3 + h_8^* r_3 + h_5^* r_1 - h_6^* r_1$ $\left.+h_7^* r_1 - h_8^* r_1\right) - s_3\left.\right\|^2 + \left(-1 + \sum_{i=1}^{8}|h_i|^2\right)|s_3|^2\left.\right]\right\}$
s_4	$Arg\left\{\min_{s_4 \in S}\left[\left\|\frac{1}{2\sqrt{2}}\left(2h_4 r_5^* - h_5 r_1^* + 2h_2 r_5^* + h_8 r_1^* - h_8 r_3^*\right.\right.\right.\right.$ $+h_6 r_3^* + h_7 r_3^* + h_6 r_4^* - h_5 r_4^* + h_6 r_1^* + h_7 r_4^* - h_8 r_4^* - h_7 r_2^* + h_6 r_2^*$ $-h_5 r_2^* + h_8 r_2^* - h_7 r_1^* + 2h_2 r_7^* - 2h_4 r_7^* - h_5 r_3^* + h_5^* r_2 + h_5^* r_4$ $+h_6^* r_4 - h_7^* r_4 - h_8^* r_4 + h_6^* r_2 + h_7^* r_2 + h_8^* r_2 + 2h_1^* r_6 + 2h_3^* r_6$ $+2h_1^* r_8 - 2h_3^* r_8 - h_5^* r_3 - h_6^* r_3 + h_7^* r_3 + h_8^* r_3 - h_5^* r_1 - h_6^* r_1$ $\left.-h_7^* r_1 - h_8^* r_1\right) - s_4\left.\right\|^2 + \left(-1 + \sum_{i=1}^{8}|h_i|^2\right)|s_4|^2\left.\right]\right\}$

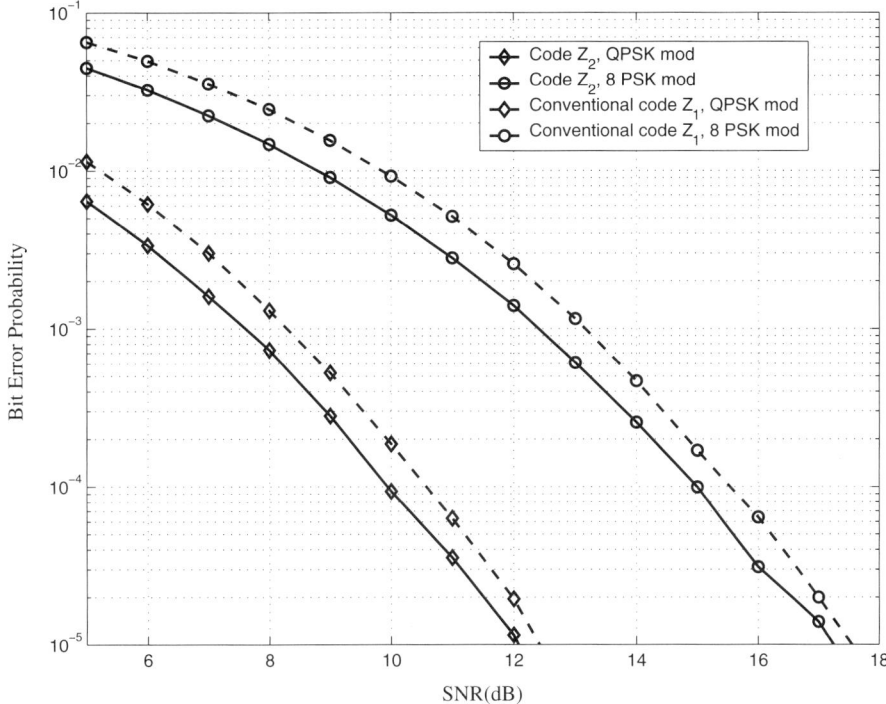

Figure 3.1. The performance of code $\mathbf{Z}_2$ compared to that of the conventional code $\mathbf{Z}_1$ in Rayleigh fading channels.

3.5 Simulation Results

In this section, the bit error performance of the proposed CO STBCs in a flat Rayleigh fading channel is presented. The channel coefficients and noise are assumed to be identically independently distributed (i.i.d.), zero-mean, complex Gaussian random variables. The signal-to-noise ratio (SNR) examined here is the channel SNR, i.e., the ratio between the sum of the average power of all received signals during a symbol time slot (STS) at the receiver antenna and the average noise power. In the simulations, we examine the performance of the proposed CO STBCs $\mathbf{Z}_2$, $\mathbf{Z}_3$ and $\mathbf{Z}_4$ in comparison with that of the conventional code $\mathbf{Z}_1$, where numerous zero entries are contained in the code matrix, in both QPSK and 8PSK modulation schemes.

Note that, in our simulations, the *power transmitted through each transmitter antenna in each STS* is normalized to one in both QPSK and 8PSK modulation cases. In particular, for $\mathbf{Z}_2$, the transmitted symbols s_1 and s_2 are derived from

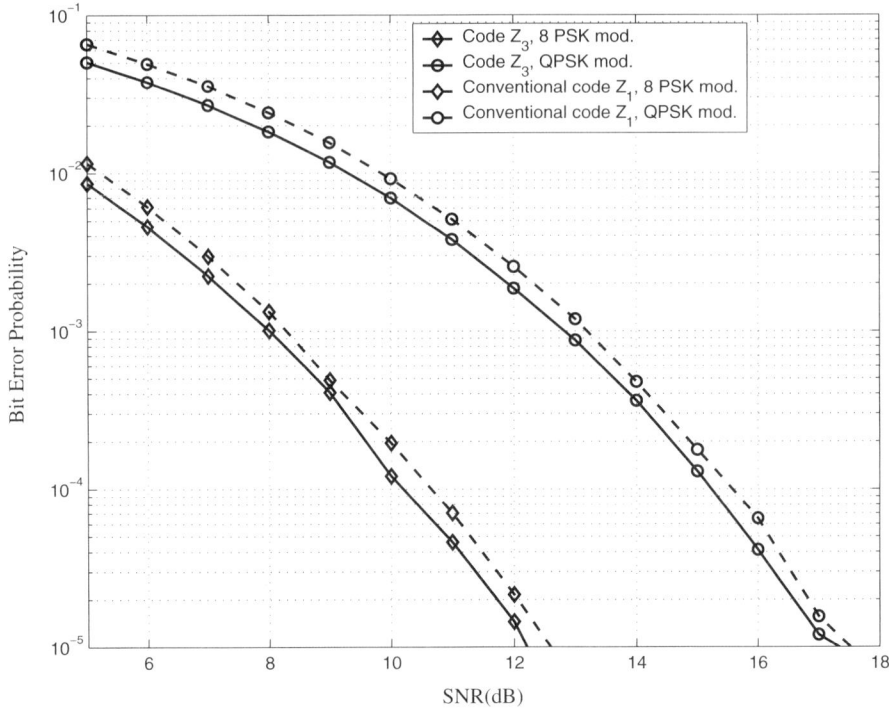

Figure 3.2. The performance of code $\mathbf{Z}_3$ compared to that of the conventional code $\mathbf{Z}_1$ in Rayleigh fading channels.

a signal constellation of power 1, while s_3 and s_4 are derived from a signal constellation of power 2. Similarly, for $\mathbf{Z}_3$, the transmitted symbols s_1, s_2 and s_3 are derived from a signal constellation of power 1, while s_4 is derived from a signal constellation of power 4. For $\mathbf{Z}_4$, all the transmitted symbols are derived from a signal constellation of power 2, while for the conventional code $\mathbf{Z}_1$, they are derived from a constellation of power 1.

Clearly, the *peak* transmission power for each symbol in the whole block of $\mathbf{Z}_2$, $\mathbf{Z}_3$ or $\mathbf{Z}_4$ is at least equal to that in the conventional code $\mathbf{Z}_1$. Therefore, the simulation results are expected to show that the performance of the proposed codes is better than that of $\mathbf{Z}_1$. The main purpose of presenting the simulations here is to show how significantly the bit error performance of the proposed codes are better than that of the conventional code, while we stick to the aim of transmitting the power of information-bearing symbols equally through each Tx antenna per STS, which is, in turn, the main motive of proposing these CO STBCs.

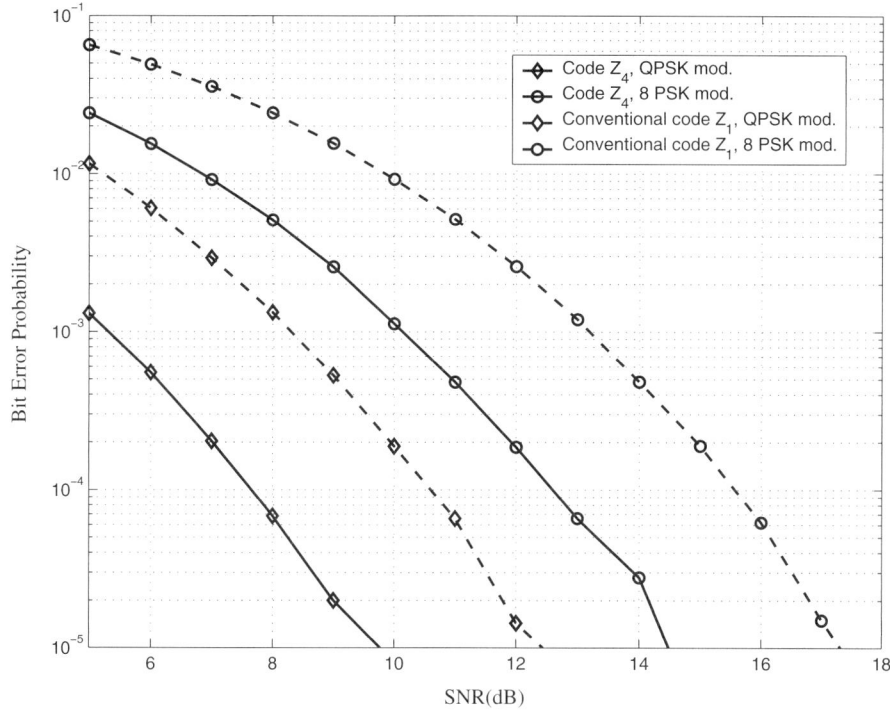

Figure 3.3. The performance of code $\mathbf{Z}_4$ compared to that of the conventional code $\mathbf{Z}_1$ in Rayleigh fading channels.

Figures 3.2 and 3.1 show that the proposed CO STBCs $\mathbf{Z}_3$ and $\mathbf{Z}_2$ provide better bit error performance than the conventional code by approximately 0.5 dB and 1 dB at the bit error rate $BER = 10^{-2}$, respectively, in both QPSK modulation and 8PSK modulation. The fact that $\mathbf{Z}_2$ provides a better bit error property than $\mathbf{Z}_3$ can be intuitively interpreted as follows. In $\mathbf{Z}_3$, only s_4 is repeated multiple times in both temporal and spatial directions, while, in $\mathbf{Z}_2$, both s_3 and s_4 are repeated. In addition, it can be observed that the symbols are *better scattered* in the code block of $\mathbf{Z}_2$ than the symbols in $\mathbf{Z}_3$. In other words, code $\mathbf{Z}_2$ provides more diversity (temporal and spatial) for the bits embedded in the symbols s_3 and s_4, while $\mathbf{Z}_3$ provides more diversity only for the bits embedded in s_4. Therefore, $\mathbf{Z}_2$ has a better resistance to burst errors than $\mathbf{Z}_3$.

It is clearly presented by Fig. 3.3 that the proposed CO STBC $\mathbf{Z}_4$ is superior to the conventional one (and $\mathbf{Z}_2$, $\mathbf{Z}_3$ as well). Particularly, at $BER = 10^{-4}$, it provides a 2.5 dB better bit error performance than the conventional code in both QPSK and 8PSK modulation schemes.

3.6   Conclusions

In this chapter, we have presented three new *square, maximum rate,* order-8 CO STBCs based on $COD(8; 1, 1, 2, 2)$, $COD(8; 1, 1, 1, 4)$ and $COD(8; 2, 2, 2, 2)$ where fewer time slots are wasted, compared to the conventional code. As a result, the new codes require lower peak-to-mean power ratios at Tx antennas to achieve the same bit error performance as the conventional code Z_1. Equivalently, with the same peak power at Tx antennas, our proposed codes provide a better bit error performance than the conventional CO STBCs.

Since, some of the variables in the codes Z_2 and Z_3 appear more often than the others, these codes can be utilized for multi-modulation systems, where the variables appearing more frequently carry signals from higher order constellations than those appearing just once in each column. Those multi-modulation schemes will be mentioned in the next chapter. Moreover, because of the lack of zeros in the design Z_4, this code is much easier to implement.

Further, we may have following important observations. From the *practical point of view*, besides the rank and determinant (or coding advantage) criteria [Tarokh et al., 1999b], [Tarokh et al., 1999c], [Tarokh et al., 1998], the two following *intuitive signs* should be considered in designing a good CO STBC, especially in a high data rate wireless system. First, the CO STBC should contain *as few zero entries as possible*. Second, a better dispersion of the transmitted symbols in both spatial (i.e., transmitter antennas) and temporal directions *more likely (but not necessarily)* results in a better bit error performance. Mathematically, each indeterminate in the code matrix should be *as much scattered in the whole matrix as possible*.

Chapter 4

MULTI-MODULATION SCHEMES TO ACHIEVE HIGHER DATA RATE

4.1   Introduction

It is well known that Complex Orthogonal Space-Time Block Codes (CO STBCs), for more than two transmitter (Tx) antennas with full diversity, cannot provide a full rate [Liang, 2003], [Liang and Xia, 2003], [Wang and Xia, 2003]. For instance, the maximum rates of *square* CO STBCs for 4 and 8 Tx antennas are 3/4 and 1/2, respectively.

The emphasis on the maximum achievable rate of *square*, order-8 CO STBCs being 1/2, rather than 5/8, has been mentioned in Chapter 2 of the book. Readers may refer to Remark 2.3.8 in Chapter 2 for more details.

Though the maximum rate of *square* CO STBCs may be smaller than that of *non-square* CO STBCs, *square* CO STBCs have a great advantage over *non-square* CO STBCs in that they require a much smaller length of the codes, i.e., much shorter processing delay. This has been clearly analyzed in Example 2.3.1 and in Section 3.2.

In addition, although CO STBCs for more than two Tx antennas cannot provide a full rate, they are still attractive in the sense that they can provide full spatial diversity for a given number of Tx antennas.

As analyzed in Chapter 3, the existing CO STBCs for 8 Tx antennas comprise various unused Symbol Time Slots (STSs), where no useful information is transmitted. In particular, 50% of STSs are unused in the conventional code, which is mentioned in [Tirkkonen and Hottinen, 2002] or in Eq. (2.36) in Chapter 2 of this book. The number of unused STSs in CO STBCs should be limited, since, during those slots, the Tx antennas must be turned off. This is inconvenient from the technical point of view, especially for systems transmitting a high data rate with very short symbol time slots.

In Chapter 3 (also in [Seberry et al., 2004], [Tran et al., 2004a], [Zhao et al., 2005a]), we have proposed three new *square, maximum rate*, order-8 CO STBCs based on the Amicable Orthogonal Designs (AODs) theory for 8 Tx antennas. Among them, we consider here the two codes, namely C_1 and C_2, which are given later in Fig. 4.2 (pp. 80) of this chapter. In these proposed codes, the number of unused symbol periods is only 25% and 12.5%, respectively, compared to 50% in the conventional designs with numerous unused slots. Consequently, a *smaller* peak-to-average power ratio at each Tx antenna is required to achieve the same bit error rates as in the conventional CO STBCs of the same order. This is the main advantage of our new codes over the conventional ones. In addition, limiting the number of unused STSs results in providing more space and STSs for transmitting bits, i.e., providing more spatial and temporal diversity for those bits, and consequently, providing better bit error performance than the conventional codes. This will be clearly shown in the simulation results presented later in this chapter.

In this chapter, we will take advantage of the property that, some symbols in the proposed codes appear more often than the others, in order to increase the data rate by utilizing higher level modulation schemes with higher transmission power for those symbols appearing more frequently in the codes. We refer to these techniques as multi-modulation schemes (MMSs).

Moreover, we examine the optimal inter-symbol power allocation in our proposed codes with various modulation schemes. It is tedious to generalize the MMSs increasing the rate of CO STBCs and the optimal power allocation in multi-modulated CO STBCs mentioned in this chapter to apply to CO STBCs of other orders, and therefore, we do not carry out this task here.

It is important to note that, MMSs for CO STBCs have been somewhat mentioned in [Tirkkonen and Hottinen, 2001]. However, the approach mentioned there has a shortcoming analyzed as follows (we use the notation mentioned there for ease of exposition). The bit error rate E_{QQ8} mentioned in Eq. (9) in [Tirkkonen and Hottinen, 2001] (respectively, E_{QQ16} in Eq. (13)) of the multi-modulated, order-4 CO STBC is a function of:

- g_{ave} : the average E_b/N_0 of all transmitted symbols, including single modulated and multi-modulated symbols.

- r : the ratio between the E_b/N_0, say g_8 (respectively, g_{16}), of the multi-modulated symbols and the E_b/N_0, say g_Q, of the single modulated symbols.

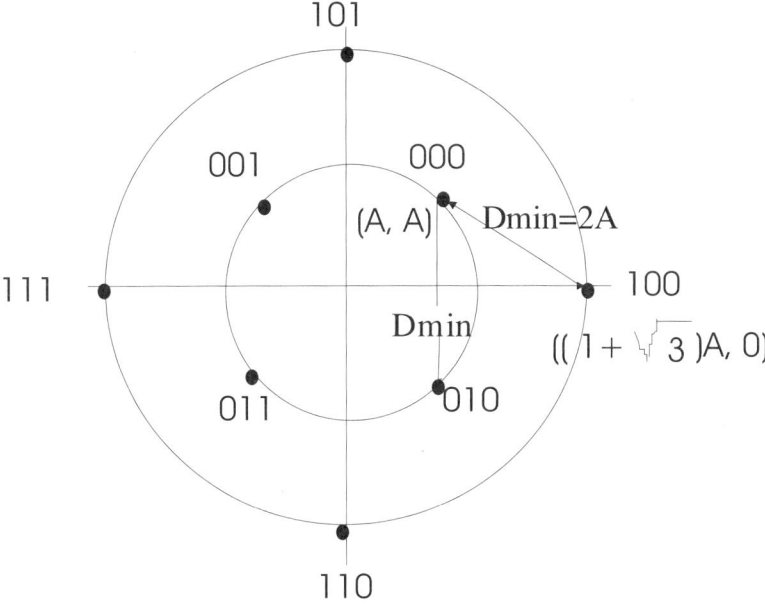

Figure 4.1. 8QAM signal constellation and bit mapping scheme.

Fig. 1 and Fig. 2 in [Tirkkonen and Hottinen, 2001] showed that, for a certain value of g_{ave}, a certain optimal value of r, say r_{opt}, is found. Equivalently, for a certain value of g_{ave}, the uniquely optimal values of g_Q and g_8 (respectively, g_Q and g_{16}) exist and these values are different for different values of g_{ave}. It means that, to achieve the optimal bit error rate, the powers of transmitted symbols (single modulated and multi-modulated symbols) must change corresponding to the change in g_{ave} (due to the change in noise affecting the transmission channels for instance).

Clearly, the approach in [Tirkkonen and Hottinen, 2001] has marginal meaning from the practical point of view, since, the average power of transmitted symbols is normally constant during the transmission, rather than changing all the time to achieve the optimal bit error rate. This is the main disadvantage of the approach mentioned in [Tirkkonen and Hottinen, 2001]. We consider the optimal value r_{opt} in this approach as the *'locally'* optimal value because this optimal value changes all the time depending on the instantaneous value of g_{ave}, except when g_{ave} is very large (about 18 dB as shown in Fig. 1 and 2 in [Tirkkonen and Hottinen, 2001]). The main reason of this shortcoming is the authors in [Tirkkonen and Hottinen, 2001] considered r_{opt} with respect to (w.r.t.) a certain value of g_{ave}.

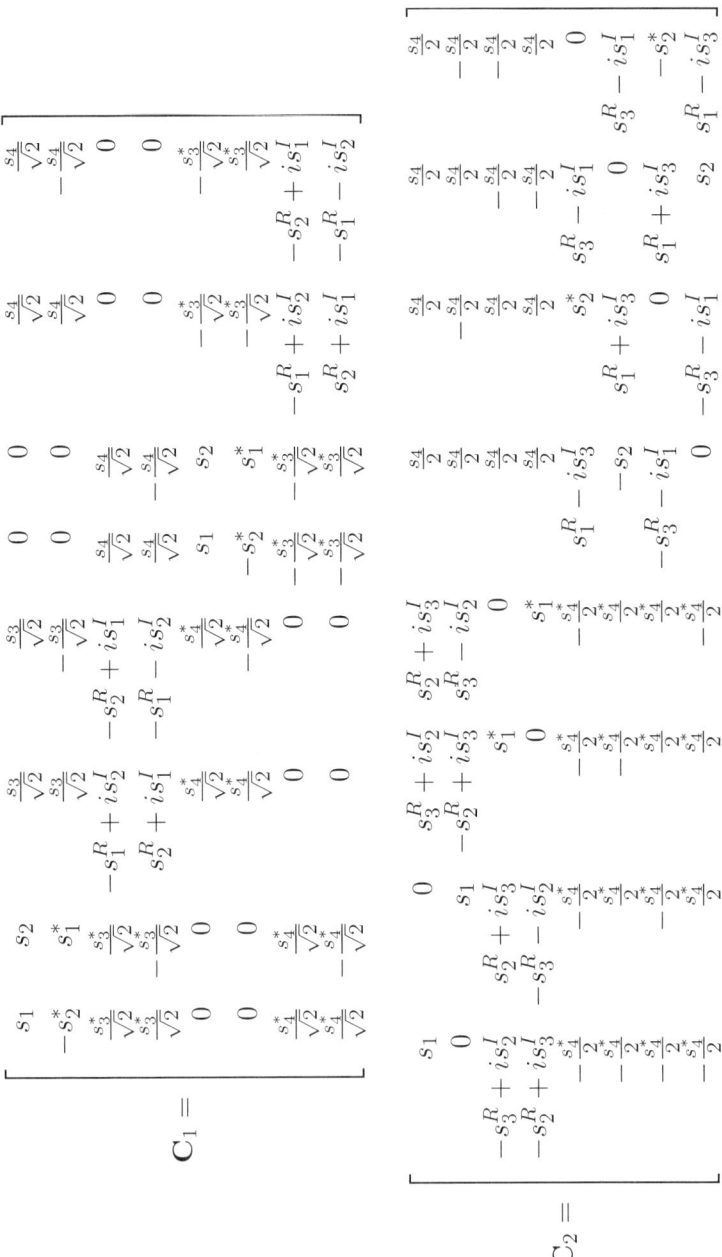

Figure 4.2.   Two new CO STBCs proposed for 8 Tx antennas.

More practical issues which could be raised are:

- Whether there exists an approach to find the optimal inter-symbol power ratio r_{opt} (between the bit SNR E_b/N_0 of multi-modulated symbols g_8 (respectively, g_{16}) and E_b/N_0 of single modulated symbols g_Q) for a certain E_b/N_0 of *single modulated symbols* g_Q, to achieve the optimal (or almost optimal) error property. The term *'almost optimal'* here means that any further improvement in bit error property, if any, is negligible.

- Whether the optimal value r_{opt} is (almost) unchanged for a wide range of g_Q.

The solution for the aforementioned issues, if any, is a more practical approach, since, for the whole range of E_b/N_0 of the single modulated symbols, we can select the inter-symbol power ratio r to be equal to r_{opt}, which is constant, to achieve the optimal error property with the consequence that the transmitted power per symbol as well as per Tx antenna does not change all the time. Clearly, in this approach, the uniquely optimal value r_{opt} is found for a whole range of g_Q, rather than changing depending on the instantaneous value of g_{ave} as mentioned in [Tirkkonen and Hottinen, 2001].

This chapter proposes such a practical approach. As stated later in this chapter, we search for the optimal inter-symbol power ratio r_{opt} w.r.t. E_b/N_0 of the single modulated symbols (as opposed to the average E_b/N_0 g_{ave} like in [Tirkkonen and Hottinen, 2001]), i.e., we search for the optimal power of the higher level modulated symbols corresponding to a given power of the single modulated symbols. This optimal ratio r_{opt} stays constant for a wide range of the signal-to-noise ratio per bit (or the bit SNR) of the single modulated symbols. Therefore, the optimal value r_{opt} in our approach may be considered as the *'globally'* optimal value, rather than being a *'locally'* optimal value as in [Tirkkonen and Hottinen, 2001].

The content of this chapter is based on the following published papers [Tran et al., 2004d], [Tran et al., 2004e].

The chapter is organized as follows. In Section 4.2, the two new STBCs proposed for 8 Tx antennas in Chapter 3 are recalled for ease of exposition. In the next section, the MMSs increasing the rate of the proposed codes are examined. The optimal ratios of symbol power in different modulation schemes for $\mathbf{C}_1$ and $\mathbf{C}_2$ in both Additive White Gaussian Noise (AWGN) and flat Rayleigh fading channels are examined in Section 4.4. Simulation results are presented in Section 4.5 and the chapter is concluded by Section 4.6. The formulas for symbol error probability of M-ary PSK signals in flat Rayleigh fading channels are derived in Appendix A.

4.2   Two New Complex Orthogonal STBCs For Eight Transmitter Antennas

In this chapter, we consider the two new, *square*, order-8 CO STBCs proposed in Chapter 3. For ease of exposition, we recall these CO STBCs as given in Fig. 4.2. It is clear that, with single modulation, the proposed codes provide the *maximum* code rate of 1/2 as in the case of the conventional codes [Tarokh et al., 1999b], [Tirkkonen and Hottinen, 2002].

4.3   MMSs to Increase the Data Rate

It is evident that, in C_1 and C_2, some symbols are transmitted in more than a single time slot per given antenna. In fact, in C_1, symbols s_3 and s_4 are transmitted twice as often as s_1 or s_2. In C_2, the symbol s_4 is transmitted four times as often as s_1, s_2 or s_3. Thus, by associating s_3 and s_4 in C_1 and s_4 in C_2 with symbols from multilevel complex modulation schemes and the remaining symbols in each of C_1 and C_2 with QPSK symbols, the overall code rates can be increased (there is, certainly, a tradeoff between the rate increase and the bit error performance). These *combined* modulation schemes are referred to as Multi-Modulation Schemes (MMSs).

Particularly, if the (QPSK+8PSK) and (QPSK+16QAM) MMSs are used, then the code rate increases from 1/2 to 5/8 and 3/4 for C_1, and to 9/16 and 5/8 for C_2, respectively [Tran et al., 2004a]. The transmission power in each STS is equally allocated and normalized to 1. It means that, s_3 and s_4 in C_1 are derived from an 8PSK or 16-QAM signal constellation of power 2, while s_4 in C_2 is derived from an 8PSK or 16-QAM signal constellation of power 4. All other symbols modulated by a QPSK signal constellation in the codes have unit power.

Additionally, the MMS, which employs a QPSK signal constellation associated with an 8QAM one (see Fig 4.1) and is denoted as the (QPSK+8QAM) MMS, can be utilized to improve further the bit error performance of the proposed codes C_1 and C_2 at the same bandwidth efficiency (same code rate) as when an 8PSK (instead of 8QAM) signal constellation is used. Particularly, the symbols s_3 and s_4 in C_1 are modulated by an 8QAM constellation of power 2, while s_4 in C_2 is modulated by the one of power 4. Other symbols in the codes are derived from a unit power QPSK constellation.

The coordinates of the 8QAM signal points, presented as functions of the factor A, are given in Fig 4.1. It is easy to realize that, if the Euclidean distance between the two closest symbols in the constellation is $D_{min}=2A$ then the average transmitted signal power is $P_{av} = 4.73A^2$ [Proakis, 2001]. An 8QAM signal constellation provides a better error property than an 8PSK one, because, in order to have the same average power per symbol $P_{av} = 4.73A^2$ as in the former case, the Euclidean distance between the closest signal points in the later case d_{min} is smaller than that of the former case. Specifically, $d_{min}=1.665A$, i.e., $\frac{d_{min}}{D_{min}}=0.83$. Intuitively, the orthogonality of the signals has been partially relaxed in the 8QAM constellation to increase the Euclidean distance between the closest signal points.

It is important to distinguish the code rate of CO STBCs *without* considering modulation schemes and that of CO STBCs when modulation schemes are considered. The maximum rate of CO STBCs is bounded by Eq. (2.31) or Eq. (2.33) in Chapter 2, depending on whether the CO STBCs are *square* or *non-square*, respectively. They are the upper bounds for the code rates of CO STBCs *without* the consideration of modulation. When MMSs are considered, the code rates have *no* upper bounds. Instead, the code rates of CO STBCs in this case depend on the modulation constellations. For the M-PSK or M-QAM constellation, the rate would be different for different M.

4.4 Optimal Inter-Symbol Power Allocation in Single Modulation and in MMSs

Equal allocation of the power transmitted at each Tx antenna and in each symbol period is optimal from the transmission point of view. However, in order to make sure whether the best error performance of the codes in different modulation schemes can be achieved, the optimal power allocation between the symbols in the codes, i.e. the ratio between the power of the symbols, must be examined. We consider here both scenarios of AWGN and independent, flat Rayleigh fading channels.

4.4.1 AWGN Channels

We derive here the symbol error rates (SERs) of QPSK single modulation schemes, (QPSK+8PSK) MMSs and (QPSK+16QAM) MMSs. To do that, at first, we have the SERs of QPSK, 8PSK and 16QAM symbols in AWGN channels as follows (see (5.2-59), (5.2-61) in [Proakis, 2001] and (5.17), (5.18) and (5.19) in [Webb and Hanzo, 1994]):

$$P_{QPSK} = 2Q(\sqrt{2\gamma})[1 - 0.5Q(\sqrt{2\gamma})] \tag{4.1}$$

$$P_{8PSK} = 2Q(\sqrt{6\gamma_8}\sin\frac{\pi}{8}) \tag{4.2}$$

$$P_{16QAM} = 3Q(\sqrt{\frac{4\gamma_{16}}{5}}) + Q(3\sqrt{\frac{4\gamma_{16}}{5}}) \tag{4.3}$$

where $Q(x)=\frac{1}{\sqrt{2\pi}}\int_x^\infty e^{-\frac{t^2}{2}}dt$; $\mu=\sqrt{\frac{2\gamma}{1+2\gamma}}$; γ, γ_8 and γ_{16} are the bit SNRs of QPSK, 8PSK and 16QAM symbols, respectively.

We now calculate the $SERs$ of $\mathbf{C}_1$ and $\mathbf{C}_2$ in different modulation schemes. Let us consider the case where the symbols s_1 and s_2 in the code $\mathbf{C}_1$ are QPSK modulated, while s_3 and s_4 are 8PSK modulated as an example. It is noted that, *in each row (and column) in* $\mathbf{C}_1$, *the transmission power of the symbol* s_j, for $j = 3$ and 4, is $|s_j|^2$. It means that, *from the transmission power point of view*, only one symbol s_j is transmitted, although it may appear multiple times. Therefore, among four transmitted symbols s_1, s_2, s_3 and s_4, the probability when QPSK symbols are transmitted in $\mathbf{C}_1$ is 50%, and the probability when 8PSK symbols are transmitted is 50%. Consequently, the average SER of the code $\mathbf{C}_1$ is

$$P_{QQ88} = \frac{1}{2}P_{QPSK} + \frac{1}{2}P_{8PSK} \tag{4.4}$$

Let E_{s_k} be the average power of the symbol s_k, for $k=1\ldots4$, and r be the intersymbol power ratio of the proposed codes, which is defined as $r = \frac{E_{s_i}}{E_{s_j}}$, where $i = 3, 4$; $j = 1, 2$ for $\mathbf{C}_1$ and $i = 4$; $j = 1, \ldots, 3$ for $\mathbf{C}_2$. Clearly, in MMSs, r is the ratio between the power of the higher level modulated symbols (8PSK or 16-QAM) and that of the QPSK modulated ones. If we denote N_0 to be the variance of noise at the receiver antenna, γ_s the *average* SNR per symbol and γ_b the *average* SNR per bit, then the power ratio can be rewritten as follows:

$$r = \frac{E_{s_i}/N_0}{E_{s_j}/N_0} = \frac{\gamma_{s_i}}{\gamma_{s_j}}$$

where $\gamma_s = \gamma_b log_2 M$ for an M-ary modulated symbol. Particularly, in the (QPSK+8PSK) MMS, $r = \frac{\gamma_{s_i}}{\gamma_{s_j}}$, for $i = 3, 4$; $j = 1, 2$, i.e.

$$r = \frac{3\gamma_8}{2\gamma} \tag{4.5}$$

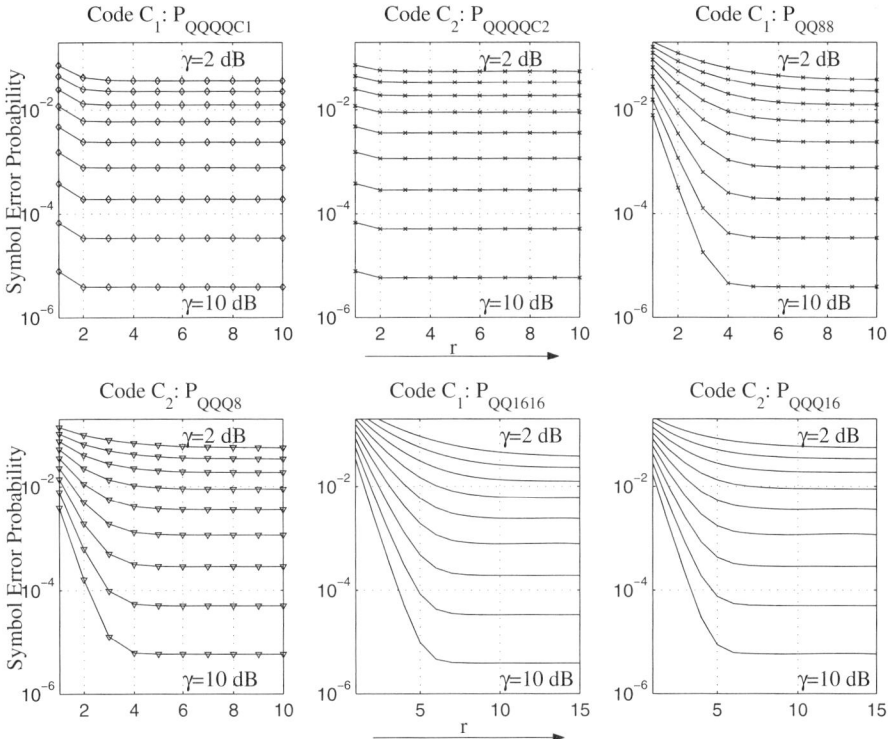

Figure 4.3. SER vs. r in single modulation and MMSs depending on γ in AWGN channels.

Therefore, if average symbol error probabilities are presented as functions of r and γ (SNR per bit of QPSK modulated symbols), then from (4.1), (4.2), (4.4) and (4.5), we have the average SER of $\mathbf{C}_1$ as given below[1]:

$$P_{QQ88} = Q(\sqrt{2\gamma})[1 - 0.5Q(\sqrt{2\gamma})] + Q[\sqrt{4r\gamma}\sin(\pi/8)] \quad (4.6)$$

Similarly, the average SER for $\mathbf{C}_2$ is

$$P_{QQQ8} = 1.5Q(\sqrt{2\gamma})[1 - 0.5Q(\sqrt{2\gamma})] + 0.5Q(\sqrt{4r\gamma}\sin(\pi/8))$$

---

[1]The channel SNR, which is used to simulate in this chapter and is defined in Section 4.5 as $SNR = \frac{\sum_{k=1}^{4} E_{s_k}}{N_0}$, is a linear function of γ (for a given value r). Additionally, the symbol error probability is a monotonically decreasing function with respect to (w.r.t.) γ for a given value r (see (4.6) for instance). Hence, if the best error performance w.r.t. γ is achieved, then that w.r.t. SNR is also achieved. Based on these notes, in the proposed method, we search for the optimal inter-symbol power ratio r_{opt} w.r.t. γ, i.e., we search for the optimal power of the higher level modulated symbols corresponding to a given power of the QPSK modulated symbols.

Table 4.1. The optimality of power allocation in single modulation and MMSs in AWGN channels.

Modulation scheme	Min. r_{opt}	
	$\mathbf{C}_1$	$\mathbf{C}_2$
Single QPSK	2	2
QPSK + 8PSK	6	4
QPSK + 16QAM	12	8

Following this method to calculate symbol error probabilities, we derive the average SERs of QPSK single modulation schemes and (QPSK+16QAM) MMSs for $\mathbf{C}_1$ and $\mathbf{C}_2$ in AWGN channels as:

- (QPSK+16QAM) multi-modulation:

$$
\begin{aligned}
P_{QQ1616} &= Q(\sqrt{2\gamma})[1 - 0.5Q(\sqrt{2\gamma})] \\
&+ 1.5Q(\sqrt{0.4r\gamma}) + 0.5Q(3\sqrt{0.4r\gamma}) \\
P_{QQQ16} &= 1.5Q(\sqrt{2\gamma})[1 - 0.5Q(\sqrt{2\gamma})] \\
&+ 0.75Q(\sqrt{0.4r\gamma}) + 0.25Q(3\sqrt{0.4r\gamma})
\end{aligned}
$$

- QPSK single modulation:

$$
\begin{aligned}
P_{QQQQC_1} &= Q(\sqrt{2\gamma})[1 - 0.5Q(\sqrt{2\gamma})] + Q(\sqrt{2r\gamma})[1 - 0.5Q(\sqrt{2r\gamma})] \\
P_{QQQQC_2} &= 1.5Q(\sqrt{2\gamma})[1 - 0.5Q(\sqrt{2\gamma})] \\
&+ 0.5Q(\sqrt{2r\gamma})[1 - 0.5Q(\sqrt{2r\gamma})]
\end{aligned}
$$

Fig. 4.3 presents the theoretical relation between SERs and r depending on γ for these modulation schemes. In this figure, γ runs from 2 dB to 10 dB, in steps of 1 dB. We can realize that, *for a given* γ, increasing r results in a better error performance when r is small. However, when r increases, the curves become flat gradually. The value at which all the curves become flat is the (smallest) optimal power allocation ratio r_{opt}. If r is selected to be higher than r_{opt}, the improvement in error property is negligible. The optimal inter-symbol power ratios r_{opt} for $\mathbf{C}_1$ and $\mathbf{C}_2$ in different modulation schemes are presented in Tables 4.1.

It is clear that the best symbol error performance can be achieved by the code $\mathbf{C}_1$ in the QPSK single modulation, since, the power ratio of this code is $r=2$, which is equal to the (smallest) optimal power ratio $r_{opt}=2$. Similarly, the best symbol error performance is also achievable by the code $\mathbf{C}_2$ in the QPSK single

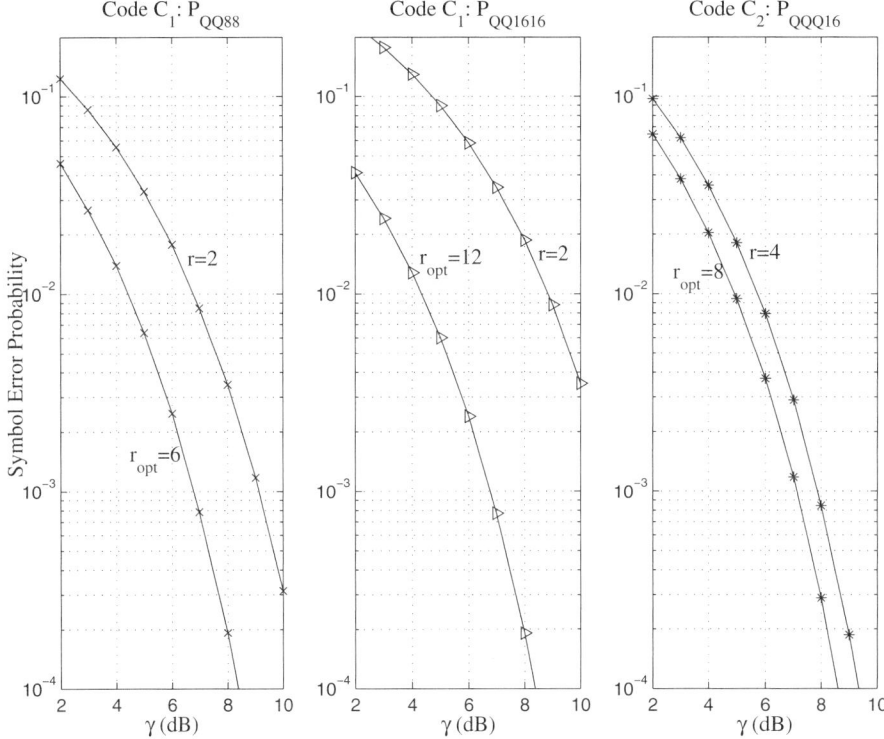

Figure 4.4. SER vs. γ with the inter-symbol power ratio $r{=}2$ for $\mathbf{C}_1$, $r{=}4$ for $\mathbf{C}_2$ and with the optimal values r_{opt} in AWGN channels.

modulation and in the (QPSK+8PSK) MMS, since, the power ratio of this code is $r{=}4$ while the (smallest) optimal power ratios are $r_{opt}{=}2$ and 4, respectively.

For the remaining modulation schemes, $r < r_{opt}$ and, consequently, there exist the significant differences between the error performance curves corresponding to r and r_{opt}, which are presented in Fig. 4.4. From Fig. 4.4, we realize that, the potential improvements for the code $\mathbf{C}_1$ in the (QPSK+8PSK) and (QPSK+16QAM) MMSs are 2.5 dB and 4.5 dB, respectively. The potential improvement for $\mathbf{C}_2$ in the (QPSK+16QAM) MMS is 0.8 dB. The potential improvement is evaluated at $SER = 10^{-2}$. The potential improvement indicates that the error performance of the proposed codes, specially for $\mathbf{C}_1$ in the (QPSK+16QAM) MMS, can be much more improved by selecting r close to r_{opt} with the penalty of unbalanced power transmission per symbol time slot at a given transmitter antenna.

In addition, it is observed from Fig. 4.3 that, in the same MMS, the code C_1 may provide a higher code rate with a lower error probability than C_2 for large r ($r \geq 3$ in the (QPSK+8PSK) MMS and $r \geq 5$ in the (QPSK+16QAM) MMS) at any γ in the considered range (2-10 dB). Hence, it is preferable to select C_1 if r is large enough, provided that balanced power transmission is not a necessary requirement of the system.

4.4.2 Flat Rayleigh Fading Channels

The SERs of QPSK, 8PSK and 16QAM symbols in flat Rayleigh fading channels are as follows (see (A.3), (A.5) in Appendix A of this book and (10.16), (10.35) and (10.42) in [Webb and Hanzo, 1994]):

$$P_{QPSK} \approx \frac{3}{4}\left[1 - \frac{\mu}{\sqrt{2 - \mu^2}}\right] \tag{4.7}$$

$$P_{8PSK} \approx \frac{7}{48\gamma_8 \sin^2 \frac{\pi}{8}} \tag{4.8}$$

$$P_{16QAM} \approx \frac{1}{4\gamma_{16}} \int_0^\infty [Q(\sqrt{\frac{y}{5}}) + Q(3\sqrt{\frac{y}{5}})]e^{\frac{-y}{4\gamma_{16}}} dy$$
$$+ (1 - \sqrt{\frac{2\gamma_{16}}{2\gamma_{16} + 5}}) \tag{4.9}$$

where $Q(x) = \frac{1}{\sqrt{2\pi}} \int_x^\infty e^{-\frac{t^2}{2}} dt$; $\mu = \sqrt{\frac{2\gamma}{1+2\gamma}}$; γ, γ_8 and γ_{16} are the SNRs per bit of QPSK, 8PSK and 16QAM symbols, respectively.

Similarly to the analysis mentioned in Section 4.4.1, we derive here the average SERs of C_1 and C_2 in QPSK single modulation schemes, (QPSK+8PSK) MMSs and (QPSK+16QAM) MMSs in flat Rayleigh fading channels. Particularly, if average symbol error probabilities are presented as functions of r and γ (SNR per bit of QPSK modulated symbols), then from (4.4), (4.5), (4.7) and (4.8), we have the average SER of C_1 in the (QPSK+8PSK) MMS as given below:

$$P_{QQ88} = \frac{3}{8}(1 - \sqrt{\frac{\gamma}{\gamma + 1}}) + \frac{7}{64r\gamma \sin^2 \frac{\pi}{8}}$$

The average SER for C_2 is

$$P_{QQQ8} = \frac{3}{4}P_{QPSK} + \frac{1}{4}P_{8PSK}$$
$$= \frac{9}{16}(1 - \sqrt{\frac{\gamma}{\gamma + 1}}) + \frac{7}{128r\gamma \sin^2 \frac{\pi}{8}}$$

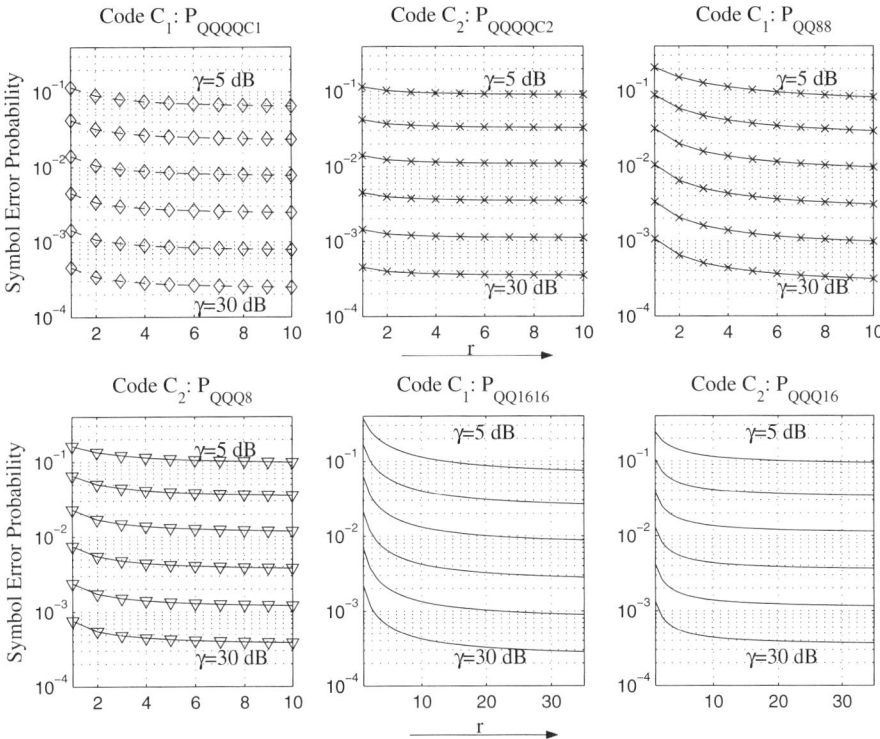

Figure 4.5. SER vs. r in single modulation and MMSs depending on γ in flat Rayleigh fading channels.

Similarly, the average SERs of $\mathbf{C}_1$ and $\mathbf{C}_2$ with QPSK single modulation schemes and with (QPSK+16QAM) MMSs in flat Rayleigh fading channels are as follows:

- (QPSK+16QAM) multi-modulation:

$$
\begin{aligned}
P_{QQ1616} &= \frac{3}{8}\left(1 - \sqrt{\frac{\gamma}{\gamma+1}}\right) + \frac{1}{2}\left(1 - \sqrt{\frac{r\gamma}{r\gamma+5}}\right) \\
&+ \frac{1}{4r\gamma}\int_0^\infty [Q(\sqrt{\frac{y}{5}}) + Q(3\sqrt{\frac{y}{5}})]e^{\frac{-y}{2r\gamma}}\,dy \\
P_{QQQ16} &= \frac{9}{16}\left(1 - \sqrt{\frac{\gamma}{\gamma+1}}\right) + \frac{1}{4}\left(1 - \sqrt{\frac{r\gamma}{r\gamma+5}}\right) \\
&+ \frac{1}{8r\gamma}\int_0^\infty [Q(\sqrt{\frac{y}{5}}) + Q(3\sqrt{\frac{y}{5}})]e^{\frac{-y}{2r\gamma}}\,dy
\end{aligned}
$$

Table 4.2.   The optimality of power allocation in single modulation and MMSs in flat Rayleigh fading channels.

Modulation scheme	Min. r_{opt}	
	$\mathbf{C}_1$	$\mathbf{C}_2$
Single QPSK	6	3
QPSK + 8PSK	8	4
QPSK + 16QAM	25	20

- QPSK single modulation:

$$P_{QQQQC1} = \frac{3}{8}(1 - \sqrt{\frac{\gamma}{\gamma+1}}) + \frac{3}{8}(1 - \sqrt{\frac{r\gamma}{r\gamma+1}})$$

$$P_{QQQQC2} = \frac{9}{16}(1 - \sqrt{\frac{\gamma}{\gamma+1}}) + \frac{3}{16}(1 - \sqrt{\frac{r\gamma}{r\gamma+1}})$$

Fig. 4.5 presents the theoretical relation between SERs and r depending on γ for these modulation schemes. In this figure, γ runs from 5 dB to 30 dB in steps of 5 dB. Similarly to the case of AWGN channels, we can realize that, *for a given value γ*, increasing r results in a better error performance when r is small. However, when r increases, the curves become flat gradually. The value at which all the curves become flat is the (smallest) optimal power allocation ratio r_{opt}. If r is selected to be higher than r_{opt}, the improvement in error property is negligible. The optimal inter-symbol power ratios r_{opt} for $\mathbf{C}_1$ and $\mathbf{C}_2$ in different modulation schemes are presented in Tables 4.2.

It is clear that the best symbol error performance can be achieved by the code $\mathbf{C}_2$ in the QPSK single modulation and in the (QPSK+8PSK) MMS, since, the power ratio of this code is $r=4$ while the (smallest) optimal power ratios are $r_{opt}=3$ and 4, respectively.

For the remaining modulation schemes, $r<r_{opt}$ and, consequently, there exist the differences between the error performance curves corresponding to r and r_{opt}, which are presented in Fig. 4.6. From Fig. 4.6, we realize that, the potential improvement for the code $\mathbf{C}_1$ is 1 dB in both QPSK single modulation scheme and (QPSK+8PSK) MMS. The potential improvements for $\mathbf{C}_1$ and $\mathbf{C}_2$ in the (QPSK+16QAM) MMS are 6 dB and 1.75 dB, respectively. The potential improvement is evaluated at $SER = 10^{-2}$. The potential improvement indicates that the error performance of the proposed codes, specially for $\mathbf{C}_1$ in the (QPSK+16QAM) MMS, can be much more improved by selecting r close

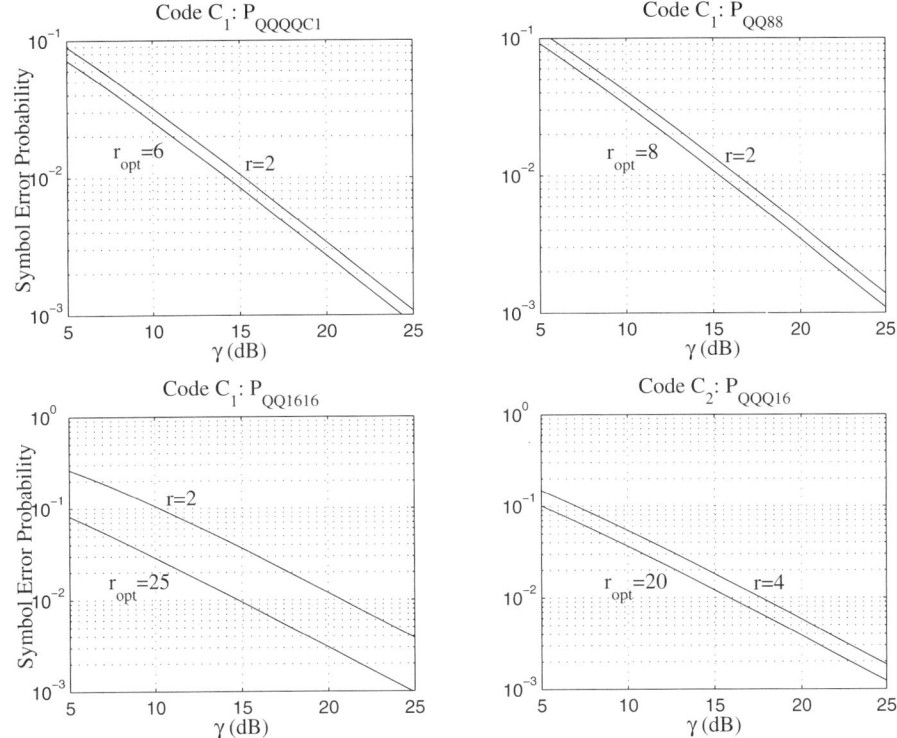

Figure 4.6. SER vs. γ with the inter-symbol power ratio $r=2$ for $\mathbf{C}_1$, $r=4$ for $\mathbf{C}_2$ and with the optimal values r_{opt} in flat Rayleigh fading channels.

to r_{opt} with the penalty of unbalanced power transmission per symbol time slot at a given transmitter antenna.

Similarly to the AWGN scenario, it is observed from Fig. 4.5 that, in the same MMS, the code $\mathbf{C}_1$ may provide a higher code rate with a lower error probability than $\mathbf{C}_2$ for large r ($r \geq 6$ in the (QPSK+8PSK) MMS and $r \geq 15$ in the (QPSK+16QAM) MMS) at any γ in the considered range (5-30 dB). Hence, it is preferable to select $\mathbf{C}_1$ if r is large enough, provided that balanced power transmission is not a necessary requirement of the system.

4.5 Simulation Results

In this section, a system comprising 8 Tx antennas and 1 Rx antenna is considered. SNR here means the channel SNR, i.e., the ratio between the total average power of the received signals and the average power of noise during each symbol time slot (STS). In all simulations, the power of signals

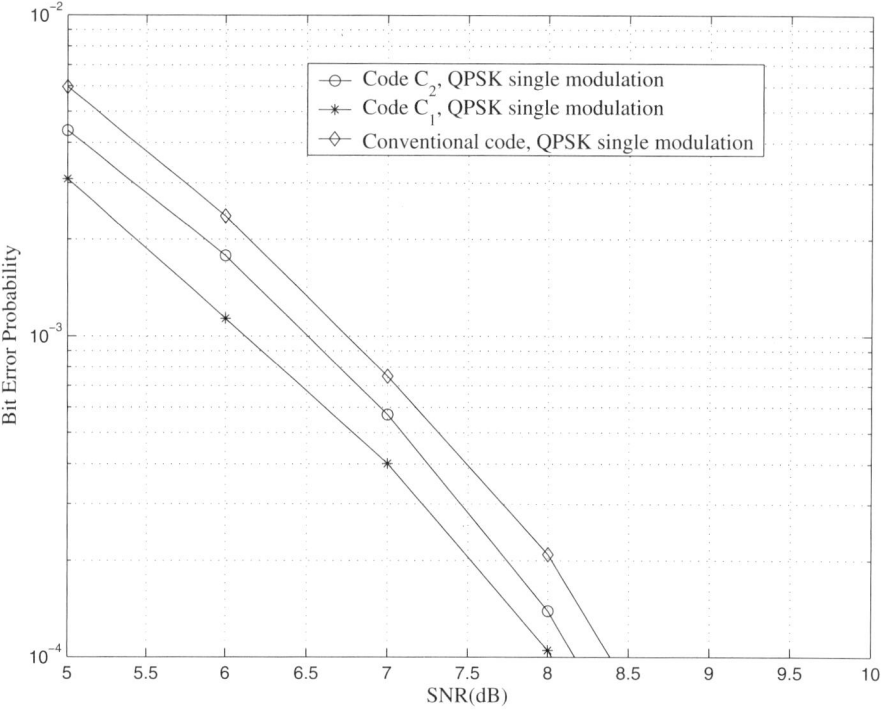

Figure 4.7. Comparison between the proposed codes and the conventional one in [Tirkkonen and Hottinen, 2002] in AWGN channels.

transmitted *in each STS* from each Tx antenna in $\mathbf{C}_1$ and $\mathbf{C}_2$ is normalized to one.

At first, the bit error properties of the codes $\mathbf{C}_1$ and $\mathbf{C}_2$ in single modulation as well as MMSs in AWGN channels are presented. Fig. 4.7 indicates that, at the bit error rate $BER=10^{-3}$, $\mathbf{C}_1$ provides 0.4 dB bit error performance better than $\mathbf{C}_2$, and 0.65 dB better than the conventional codes when QPSK single modulation is considered. The conventional codes have been mentioned in [Tirkkonen and Hottinen, 2002], or in Eq. (2.36) in Chapter 2 of this book.

As mentioned earlier in Section 3.5, the fact that $\mathbf{C}_1$ provides a better bit error property than $\mathbf{C}_2$ can be intuitively interpreted as follows. The code $\mathbf{C}_1$ provides more diversity (temporal and spatial) for 4 bits embedded in the symbols s_3 and s_4, while $\mathbf{C}_2$ provides more diversity for only 2 bits in s_4. In other words, $\mathbf{C}_1$ has a higher resistance to burst errors than $\mathbf{C}_2$. Therefore, it is preferable to select $\mathbf{C}_1$ when QPSK single modulation is utilized for 8 Tx antennas in AWGN channels.

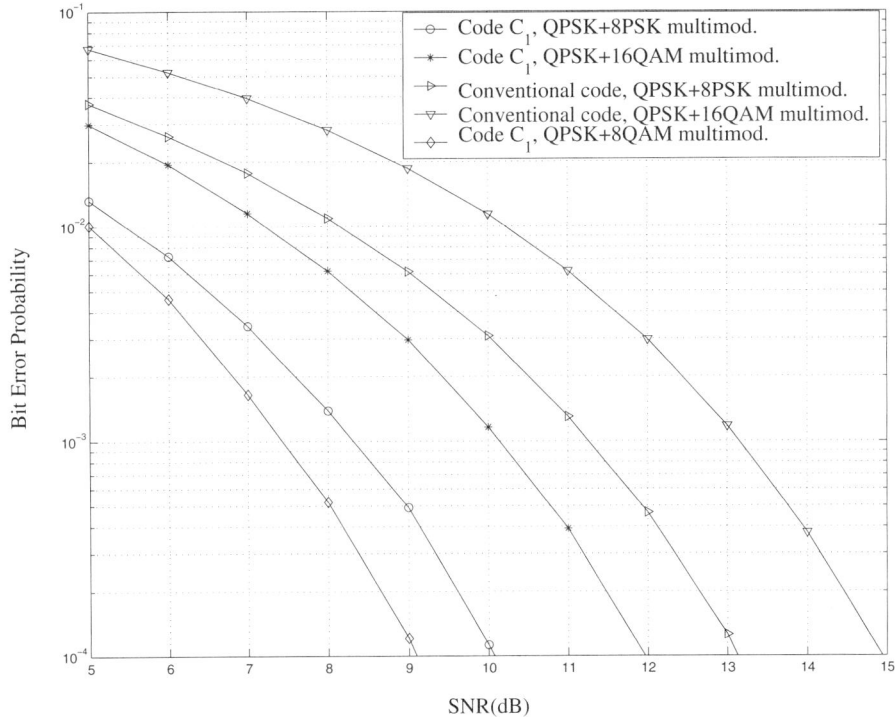

Figure 4.8. Bit error performance of the code $\mathbf{C}_1$ with different MMSs in AWGN channels.

Fig. 4.8 and Fig. 4.9 present the BERs of $\mathbf{C}_1$ and $\mathbf{C}_2$ in (QPSK+8PSK), (QPSK+8QAM) and (QPSK+16QAM) MMSs. As mentioned in Section 4.3, for the same MMS, $\mathbf{C}_1$ provides a higher code rate than $\mathbf{C}_2$. The performance of the *conventional code* with those MMSs is presented here as the reference to evaluate the superiority of our codes (evaluation must be carried out in the same MMS, i.e., at the same bandwidth efficiency). It is noted that, for the conventional code, both symbols s_3 and s_4 are 8PSK or 16QAM modulated in Fig. 4.8, while only the symbol s_4 is 8PSK or 16QAM modulated in Fig 4.9. In the *conventional code*, the power transmitted *per STS* from each Tx antenna is also normalized to one.

Clearly, the MMS using an 8QAM signal constellation provides better bit error performance than other schemes. Particularly, for the proposed codes, the SNR gains achieved by the (QPSK+8QAM) MMS are 0.15 dB for $\mathbf{C}_2$, and 1 dB for $\mathbf{C}_1$, respectively, to have the same $BER = 10^{-4}$ as in the (QPSK+8PSK) MMS. Additionally, at the same code rate, the proposed codes provide bet-

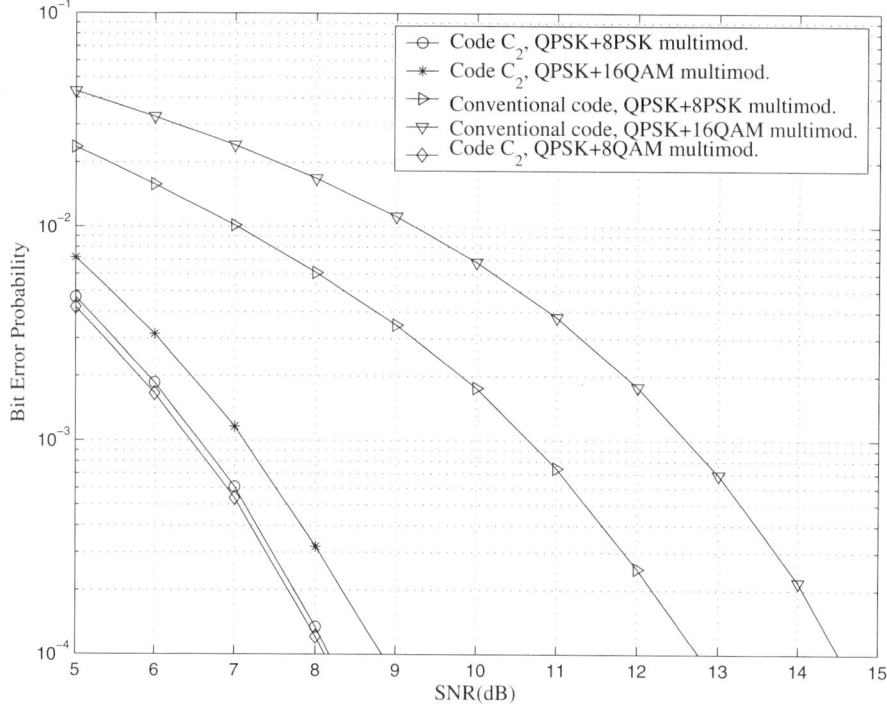

Figure 4.9.   Bit error performance of the code $\mathbf{C}_2$ with different MMSs in AWGN channels.

ter bit error performance than the *conventional code* by around 3 dB in both (QPSK+8PSK) and (QPSK+16QAM) MMSs for the case of $\mathbf{C}_1$, and around 4.5 dB in (QPSK+8PSK) MMS and 5.7 dB in the (QPSK+16QAM) MMS for the case of $\mathbf{C}_2$, respectively, at $BER=10^{-4}$. Therefore, at $BER=10^{-4}$, the SNR gains achieved by the (QPSK+8QAM) MMS are 4.65 dB for $\mathbf{C}_2$ and 4 dB for $\mathbf{C}_1$, compared to the *conventional code* with the (QPSK+8PSK) MMS.

Next, the bit error properties of the codes $\mathbf{C}_1$ and $\mathbf{C}_2$ in single modulation as well as MMSs in independent, flat Rayleigh fading channels are presented.

Fig. 4.10 indicates that, at $BER = 10^{-3}$, $\mathbf{C}_1$ provides approximately 0.4 dB bit error performance better than $\mathbf{C}_2$, and 0.65 dB better than the conventional code, when QPSK single modulation is considered. Similarly as in the case of AWGN channels, this is because $\mathbf{C}_1$ has a higher resistance to burst errors than $\mathbf{C}_2$. Therefore, it is preferable to select $\mathbf{C}_1$ when QPSK single modulation is utilized for 8 Tx antennas in Rayleigh fading channels.

Fig. 4.11 and Fig. 4.12 present the BERs of $\mathbf{C}_1$ and $\mathbf{C}_2$ in (QPSK+8PSK), (QPSK+8QAM) and (QPSK+16QAM) MMSs. Similarly, for the *conventional*

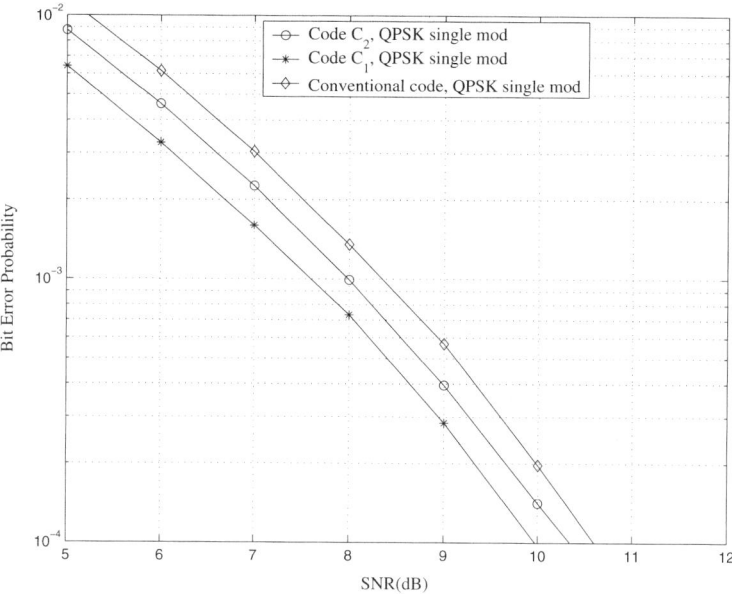

Figure 4.10. Comparison between the proposed codes and the conventional one in [Tirkkonen and Hottinen, 2002] in flat Rayleigh fading channels.

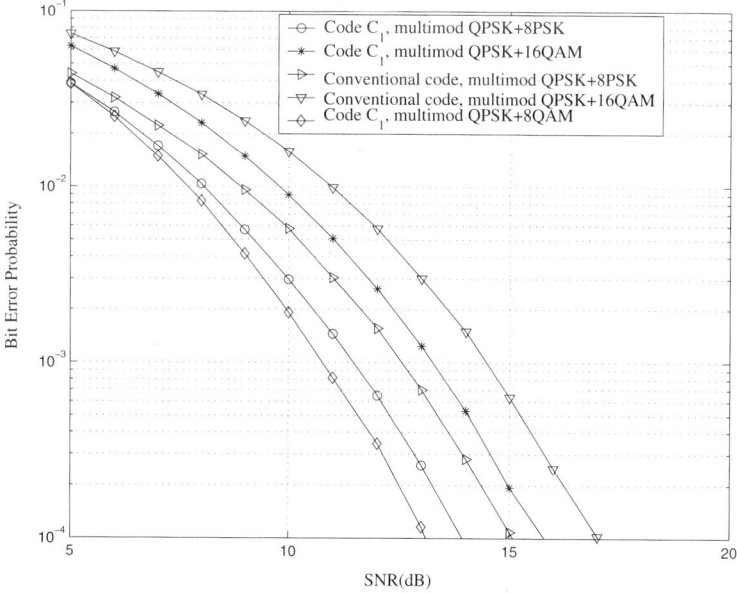

Figure 4.11. Bit error performance of the code C_1 with different MMSs in flat Rayleigh fading channels.

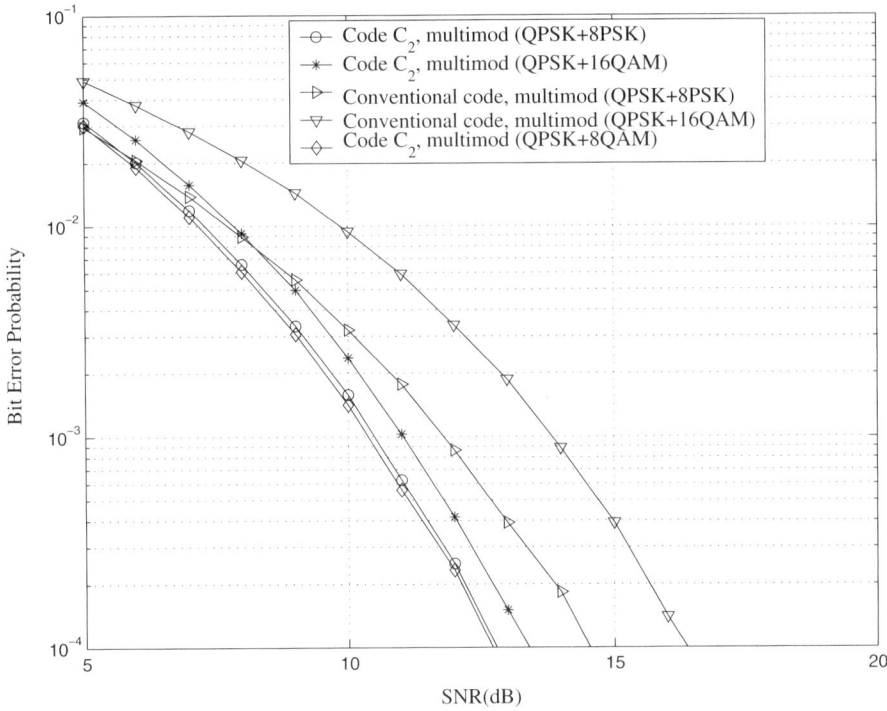

Figure 4.12. Bit error performance of the code $\mathbf{C}_2$ with different MMSs in flat Rayleigh fading channels.

code, both symbols s_3 and s_4 are 8PSK or 16QAM modulated in Fig. 4.11, while only the symbol s_4 is 8PSK or 16QAM modulated in Fig. 4.12. The power transmitted per STS from each Tx antenna is also normalized to one.

Clearly, the MMS using an 8QAM signal constellation provides better bit error performance than other schemes. Particularly, for the proposed codes, the SNR gains achieved by the (QPSK+8QAM) MMS are 0.15 dB for $\mathbf{C}_2$, and 0.8 dB for $\mathbf{C}_1$, respectively, to have the same $BER = 10^{-4}$ as in the (QPSK+8PSK) MMS. Additionally, at the same code rate, the proposed codes provide better bit error performance than the *conventional code* by around 1 dB in the (QPSK+8PSK) MMS and 1 dB in the (QPSK+16QAM) MMS for the case of $\mathbf{C}_1$, and around 1.5 dB in the (QPSK+8PSK) MMS and 2.7 dB in the (QPSK+16QAM) MMS for the case of $\mathbf{C}_2$, respectively, at $BER=10^{-4}$. Therefore, at $BER = 10^{-4}$, the SNR gains achieved by the (QPSK+8QAM) MMS are 1.65 dB for $\mathbf{C}_2$ and 1.8 dB for $\mathbf{C}_1$, compared to the *conventional code* with the (QPSK+8PSK) MMS.

4.6 Conclusions

In this chapter, MMSs have been examined to increase the rate of our proposed codes for 8 Tx antennas. In addition, a method to examine the optimal inter-symbol power allocation for the proposed codes in various modulation schemes has been derived.

It is trivial to generalize the principles of the MMSs to increase the rate of CO STBCs and of the optimal power allocation in multi-modulated CO STBCs mentioned in this chapter to apply to CO STBCs of other orders.

Our method to examine the optimal inter-symbol power ratios in MMSs overcomes the shortcoming of the conventional method mentioned in [Tirkkonen and Hottinen, 2001]. In other words, our method is more amenable to practical implementation than the method in [Tirkkonen and Hottinen, 2001].

Based on the analyses mentioned in this chapter, the following conclusions can be derived for *both* AWGN and flat Rayleigh fading channels:

1 When QPSK single modulation is utilized, it is recommended to select the code C_1, rather than C_2, for 8 Tx antennas as it provides the best BER.

2 The (QPSK+8QAM) MMS can be used to improve further the performance of the codes proposed for 8 Tx antennas, especially for C_1.

3 It turns out that selecting the power ratio $r = 4$ for C_2 is not only optimal in terms of equal power transmission per STS at each Tx antenna, but also optimal in terms of achieving the best symbol error property in the QPSK single modulation and in the (QPSK+8PSK) MMS.

4 The performance of the proposed codes can be remarkably improved, especially for the code C_1 in the (QPSK+16QAM) MMS, if the power ratio r is selected close to the optimal ratio r_{opt}, with the penalty of unbalanced power transmission per STS at a given Tx antenna.

5 In the same MMS, the code C_1 may provide a higher code rate with better error performance than C_2 if the inter-symbol power ratio is large enough, provided that balance power transmission per STS is not necessarily required for the system.

6 Finally, the principles examining the MMSs in order to increase the rate of CO STBCs and examining the optimal power allocation for multi-modulated CO STBCs mentioned in this chapter can be applied to any number of Tx

antennas, such as 4 Tx antennas as in the codes mentioned in [Tarokh et al., 1999b], [Tirkkonen and Hottinen, 2001], [Tirkkonen and Hottinen, 2002], without any difficulty. The proposed principles are useful for CO STBCs with fewer zeros in the structure, such as the codes C_1 and C_2 mentioned in this chapter, and especially useful for CO STBCs *without* any zero entries in the structure, such as the code Z_4 mentioned in Chapter 3 or the codes mentioned in our papers [Tran et al., 2004f], [Tran et al., 2005d], which will be discussed in more details in the next chapter.

Chapter 5

TWO NOVEL CONSTRUCTION CLASSES FOR IMPROVED, SQUARE CO STBCS

5.1   Introduction

A $p \times n$ CO STBC over k variables is corresponding to n transmitter (Tx) antennas, decoding delay or memory length of p, and rate $R = \frac{k}{p}$. Given n and R, the goal is to minimize the decoding delay p. Hence, *square* CO STBCs are particularly interesting because they require the minimum processing delay (minimum memory length) for a given rate and a given number of Tx antennas.

Another consideration for practical implementation is the number of zero entries in a code. Compared to a code with fewer zeros, a code with more zeros results in a higher peak-to-mean power ratio for the Tx antennas to achieve the same Bit Error Rate (BER). Having many zeros can also impede practical implementation since some Tx antennas must be turned off during transmission. Turning Tx antennas off during transmission is inconvenient, especially for high data rate wireless communication systems.

Furthermore, it would be more practical if the power of signals could be equally transmitted via each Tx antenna during every symbol time slot (STS).

Given the above considerations for CO STBCs, this chapter focuses on constructing *square* CO STBCs with *maximum rates*, *minimum* decoding delays, *no* zero entries, and *equal* power transmission per Tx antenna during each STS. We refer the *square* CO STBCs with those properties to as the *improved, square* CO STBCs.

The simplest *square* CO STBC is the Alamouti code [Alamouti, 1998], which achieves a rate of one for two Tx antennas. In contrast, *square* CO STBCs for more than two Tx antennas cannot achieve the rate of one (see Sections 2.3.2 in Chapter 2 of this book, or [Liang, 2003], [Liang and Xia, 2003]), but they can still achieve full diversity for a given number of Tx antennas. Constructions of *square* CO STBCs for a higher number of Tx antennas, e.g. 4 and 8, have been well examined in the literature, such as in [Tirkkonen and Hottinen, 2002] and [Liang, 2003]. As mentioned earlier in Sections 2.3.2 of this book, these structures yield *square* CO STBCs of the *maximum rate*, which is, for instance, 1/2 for 8 Tx antennas. However, these maximum rate codes have many zero entries, which are undesirable.

Square CO STBCs with no zero entries have been proposed in the literature, such as [Alamouti, 1998] and [Tarokh et al., 1999b], for orders 2 and 4. In [Tran et al., 2004a], we constructed two square, order-8 CO STBCs with *fewer zeros* than the conventional codes [Liang, 2003], [Tirkkonen and Hottinen, 2002]. These codes have also been mentioned in our published works [Seberry et al., 2004], [Zhao et al., 2005a] and in Chapters 3 and 4 of this book. Further, in [Seberry et al., 2004] and [Zhao et al., 2005a], we constructed a *square, maximum rate*, order-8 CO STBC **Z** *without any zero entries*:

$$
\mathbf{Z} = \begin{bmatrix}
s_1 & s_1 & s_2 & s_2 & s_3 & s_4 & s_3 & s_4 \\
s_1 & -s_1 & s_2 & -s_2 & s_4^* & -s_3^* & s_4^* & -s_3^* \\
s_2^* & s_2^* & -s_1^* & -s_1^* & s_3 & s_4 & -s_3 & -s_4 \\
s_2^* & -s_2^* & -s_1^* & s_1^* & s_4^* & -s_3^* & -s_4^* & s_3^* \\
-s_4^R+is_3^I & -s_3^R+is_4^I & -s_4^R+is_3^I & -s_3^R+is_4^I & s_2^R-is_1^I & s_2^R-is_1^I & s_1^R-is_2^I & s_1^R-is_2^I \\
-s_3^R-is_4^I & s_4^R+is_3^I & -s_3^R-is_4^I & s_4^R+is_3^I & s_2^R-is_1^I & -s_2^R+is_1^I & s_1^R-is_2^I & -s_1^R+is_2^I \\
-s_4^R+is_3^I & -s_3^R+is_4^I & s_4^R-is_3^I & s_3^R-is_4^I & s_1^R+is_2^I & s_1^R+is_2^I & -s_2^R-is_1^I & -s_2^R-is_1^I \\
-s_3^R-is_4^I & s_4^R+is_3^I & s_3^R+is_4^I & -s_4^R-is_3^I & s_1^R+is_2^I & -s_1^R-is_2^I & -s_2^R-is_1^I & s_2^R+is_1^I
\end{bmatrix}
$$

This new CO STBC has also been mentioned in Chapter 3 of this book.

As pointed out in [Seberry et al., 2004] and in Chapter 3 of this book, the entries z_{lk} ($l = 5, \ldots, 8$, $k = 1, \ldots, 8$) of $\mathbf{Z}$ are composed of the real part of one indeterminate and the imaginary part of another indeterminate, e.g., $z_{51} = -s_4^R + i s_3^I$. This observation means that if the indeterminates $s_1, \ldots, s_4$ are chosen from the complex signal constellations where s_j^R or s_j^I ($j=1\ldots4$) can be equal to zero, e.g., the QPSK constellation $(1,-1,i,-i)$ then, some of the entries of the matrix $\mathbf{Z}$ can be equal to zero depending on the transmitted data. Therefore, such constellations should be avoided. An example of the constellation where the power is evenly spread among the Tx antennas independently of the transmitted data is the QPSK constellation $(1+i,1-i,-1+i,-1-i)$.

The proposed square CO STBC $\mathbf{Z}$ has the following advantages:

1 It is not required to turn off any Tx antenna during transmission, unlike in the conventional CO STBC [Liang, 2003], [Tirkkonen and Hottinen, 2002].

2 When the indeterminates are chosen from a suitable constellation, $\mathbf{Z}$ has no zero entries, hence, it requires a smaller peak power per Tx antenna to achieve the same BER as in the conventional square CO STBCs with zeros [Liang, 2003], [Tirkkonen and Hottinen, 2002]. Equivalently, it provides a better BER compared to the conventional square CO STBCs with the same peak power at Tx antennas.

Independently, from *Amicable Orthogonal Designs* (AODs), C. Yuen et al. [Yuen et al., 2004] constructed the following *solitary*, square, order-8 CO STBC with no zeros:

$$
\mathbf{G}8 = \begin{bmatrix}
s_1^* & s_1^* & s_2 & -s_2 & s_3 & -s_3 & s_4 & -s_4 \\
js_1 & -js_1 & js_2^* & js_2^* & js_3^* & js_3^* & js_4^* & js_4^* \\
-s_2 & s_2 & s_1^* & s_1^* & s_4^* & -s_4^* & -s_3^* & s_3^* \\
-js_2^* & -js_2^* & js_1 & -js_1 & js_4 & js_4 & -js_3 & -js_3 \\
-s_3 & s_3 & -s_4^* & s_4^* & s_1^* & s_1^* & s_2^* & -s_2^* \\
-js_3^* & -js_3^* & -js_4 & -js_4 & js_1 & -js_1 & js_2 & js_2 \\
-s_4 & s_4 & s_3^* & -s_3^* & -s_2^* & s_2^* & s_1^* & s_1^* \\
-js_4^* & -js_4^* & js_3 & js_3 & -js_2 & -js_2 & js_1 & -js_1
\end{bmatrix} \tag{5.1}
$$

where $j = \sqrt{-1}$. The background knowledge on AODs can be found in [Geramita and Seberry, 1979]. This square CO STBC has an advantage over our code $\mathbf{Z}$ in that it does not require the restriction on signal constellations. However, from amicable orthogonal designs, it is difficult to construct square

CO STBCs, especially for those codes of high orders, since we have to incorporate many weighting matrices. For instance, to construct a square, *maximum rate* CO STBC of order 8, we have to find 8 matrices of size 8×8 (4 weighting matrices for the real parts of variables and 4 other weighting matrices for the imaginary parts) which simultaneously satisfy several strong conditions of AODs [Ganesan and Stoica, 2000], [Geramita and Seberry, 1979], [Yuen et al., 2004].

By modifying the Williamson and Wallis-Whiteman arrays to apply to complex matrices, we propose two *novel* methods of construction of *square*, order-$4n$ CO STBCs from *square*, order-n codes which satisfy certain properties. Applying the proposed methods, we construct *square*, *maximum rate*, order-8 CO STBCs with no zeros, such that the transmitted symbols equally disperse through Tx antennas. Besides having the maximum rate, the minimal decoding delay, and no zero entries, the resultant codes, referred to as the *improved square CO STBCs*, have the following practical advantages:

1 They do not require any restriction on allowable signal constellations;

2 It is possible to transmit symbols with equal power for any STS at any Tx antenna;

3 A lower peak power per Tx antenna is required to achieve the same bit error rates as for the conventional CO STBCs with zeros.

As mentioned in more details later in this chapter, in order to construct, for instance, 8×8 CO STBCs, the *main* task in our methods is to find two submatrices of size 2×2 which satisfy certain properties, rather than finding 8 weighting matrices of size 8×8 simultaneously as in the AOD approaches, such as in [Yuen et al., 2004]. More importantly, our methods give a transition from square, order-n CO STBCs satisfying certain properties to square, order-$4n$ CO STBCs. Our proposed methods might even lead to the constructions of square CO STBCs of higher orders, such as 16 or 32, with fewer zeros or even without zeros.

The content of this chapter is based on our works [Tran et al., 2004f], [Tran et al., 2005d]. We mark the discovery of the two novel constructions mentioned in this chapter by a milestone in Fig. 1.1 of Chapter 1, which presents the history of CO STBCs.

The chapter is organized as follows. In Section 5.2, we provide definitions and notations used throughout the chapter. In Section 5.3, we propose two methods for constructing high-rate, square CO STBCs of order $N = 4n$ from sub-matrices of order n. In Section 5.4, we use the proposed methods to construct square, *maximum rate*, order-8 CO STBCs, which are superior in several aspects to other known codes to date. The chapter is concluded by Section 5.5.

5.2   Definitions and Notations

Our proposed constructions in this chapter are based on the following matrices, which are the variations of the Williamson and Wallis-Whiteman arrays mentioned in [Geramita and Seberry, 1979] (pp. 121 and 99, respectively), modified to apply to complex matrices:

$$\mathcal{O}_1 = \begin{bmatrix} \mathbf{A} & \mathbf{B} & \mathbf{C} & \mathbf{D} \\ -\mathbf{B} & \mathbf{A} & \bar{\mathbf{D}} & -\bar{\mathbf{C}} \\ -\mathbf{C} & -\bar{\mathbf{D}} & \mathbf{A} & \bar{\mathbf{B}} \\ -\mathbf{D} & \bar{\mathbf{C}} & -\bar{\mathbf{B}} & \mathbf{A} \end{bmatrix} \quad (5.2)$$

$$\mathcal{O}_2 = \begin{bmatrix} \mathbf{A} & \mathbf{B} & \mathbf{C} & \mathbf{D} \\ -\bar{\mathbf{B}} & \bar{\mathbf{A}} & -\mathbf{D} & \mathbf{C} \\ -\mathbf{C} & \bar{\mathbf{D}} & \mathbf{A} & -\bar{\mathbf{B}} \\ -\bar{\mathbf{D}} & -\mathbf{C} & \mathbf{B} & \bar{\mathbf{A}} \end{bmatrix} \quad (5.3)$$

where $\bar{\mathbf{X}}$ is the matrix derived from a matrix $\mathbf{X}$ by replacing all variables in $\mathbf{X}$ by their conjugates, i.e., $\bar{\mathbf{X}}=(\mathbf{X}^H)^T$. $(.)^H$ denotes the Hermitian transposition while $(.)^T$ denotes the transposition (but not conjugate). $\mathbf{A}$, $\mathbf{B}$, $\mathbf{C}$ and $\mathbf{D}$ are $n \times n$, square, orthogonal matrices of complex variables. Hence, $\mathcal{O}_1$ and $\mathcal{O}_2$ are $4n \times 4n$ matrices of complex variables.

Let $\mathcal{O}$ be a general notation representing either $\mathcal{O}_1$ or $\mathcal{O}_2$. Define $N=4n$ and present N as $N=2^a(2b+1)$, where a and b are integers. Let $\mu(N)$ be the maximum number of variables in $\mathcal{O}$. It is well known that the maximum number of variables in the *square* CO STBC of order N is $\mu(N) = a + 1$. Readers may refer to [Geramita and Seberry, 1979], Corollary 2 in [Liang, 2003], [Tirkkonen and Hottinen, 2002], or Eq. (2.30) in Section 2.3.2.1 of this book for more details. Let μ_A, μ_B, μ_C and μ_D be the number of variables in $\mathbf{A}$, $\mathbf{B}$, $\mathbf{C}$, and $\mathbf{D}$, respectively.

Let U and $\mathcal{I}_U$ be the set of all variables in $\mathcal{O}$ and the set of all indices of elements in U, respectively. Similarly, let

$$\begin{aligned} U_1 &= \{s_{A1}, s_{A2}, \ldots, s_{A\mu_A}\} \\ U_2 &= \{s_{B1}, s_{B2}, \ldots, s_{B\mu_B}\} \\ U_3 &= \{s_{C1}, s_{C2}, \ldots, s_{C\mu_C}\} \\ U_4 &= \{s_{D1}, s_{D2}, \ldots, s_{D\mu_D}\} \end{aligned} \quad (5.4)$$

be the sets of variables in $\mathbf{A}$, $\mathbf{B}$, $\mathbf{C}$, and $\mathbf{D}$, respectively, and let $\mathcal{I}_{U_i}$, for $i=1\ldots 4$, be the sets of indices of variables in the sub-matrices $\mathbf{A}$, $\mathbf{B}$, $\mathbf{C}$, and $\mathbf{D}$, respectively.

We require that the sub-matrices $\mathbf{A}$, $\mathbf{B}$, $\mathbf{C}$, and $\mathbf{D}$ satisfy

$$\begin{cases} \bigcup U_i = U & i = 1,\ldots,4 \\ \bigcap U_i U_j = \emptyset & i \neq j \end{cases} \tag{5.5}$$

where $\emptyset$ is the empty set.

With the condition (5.5), clearly, if $\mathcal{O}$ comprises the maximum number of variables, we have

$$\mu_A + \mu_B + \mu_C + \mu_D = \mu(N) \tag{5.6}$$

Since $\mathbf{A}$ is a matrix on variables $\{s_{A1}, s_{A2}, \ldots, s_{A\mu_A}\}$, we define the vector $\mathbf{s_A} = (s_{A1}, s_{A2}, \ldots, s_{A\mu_A})$, and write

$$\mathbf{A} \;\; = \;\; \mathbf{A(s_A)} = \mathbf{A}(s_{A1}, s_{A2}, \ldots, s_{A\mu_A}).$$

Similarly, we denote the matrices $\mathbf{B}$, $\mathbf{C}$ and $\mathbf{D}$ as

$$\begin{aligned} \mathbf{B} &= \mathbf{B(s_B)} = \mathbf{B}(s_{B1}, s_{B2}, \ldots, s_{B\mu_B}) \\ \mathbf{C} &= \mathbf{C(s_C)} = \mathbf{C}(s_{C1}, s_{C2}, \ldots, s_{C\mu_C}) \\ \mathbf{D} &= \mathbf{D(s_D)} = \mathbf{D}(s_{D1}, s_{D2}, \ldots, s_{D\mu_D}) \end{aligned} \tag{5.7}$$

For simplicity of notation, we sometimes write, for example, $\mathbf{A_{s_A}}$ to represent $\mathbf{A(s_A)}$.

Recall that the matrix $\bar{\mathbf{A}}$ is derived from $\mathbf{A}$ by replacing each variable s_{Ai}, for $1 \leq i \leq \mu_A$, by its conjugate. Denote the conjugate of each variable by s_{Ai}^*, for $1 \leq i \leq \mu_A$, and denote the vector of these conjugates by $\mathbf{s_A}^*$. Now, we write

$$\bar{\mathbf{A}} \;\; = \;\; \mathbf{A(s_A}^*) = \mathbf{A}(s_{A1}^*, s_{A2}^*, \ldots, s_{A\mu_A}^*)$$

and similarly for $\bar{\mathbf{B}}$, $\bar{\mathbf{C}}$, and $\bar{\mathbf{D}}$.

We state that a matrix $\mathbf{X(s_X)}$ is *of similar form* to a matrix $\mathbf{Y(s_Y)}$ (or just $\mathbf{X}$ is *of similar form* to $\mathbf{Y}$, for short) if $\mathbf{X} = k_X \mathbf{Y(s_X)}$, where $\mathbf{s_X}$ is a vector containing distinct complex variables $s_{X1}, s_{X2}, \ldots, s_{X\mu_X}$, and similarly, $\mathbf{s_Y}$ is a vector containing distinct complex variables $s_{Y1}, s_{Y2}, \ldots, s_{Y\mu_Y}$, and k_X is an arbitrary, non-zero, real coefficient. In this notation, we stipulate that the number of variables μ_X in $\mathbf{X}$ is at most equal to the number of variables μ_Y

in $\mathbf{Y}$. To illustrate an example with $\mu_X = \mu_Y = 2$, $\mathbf{X}(\mathbf{s_X}) = \begin{bmatrix} s_{X1} & s_{X2} \\ -s_{X2}^* & s_{X1}^* \end{bmatrix}$ (which presents the Alamouti code with two variables) is of similar form to $\mathbf{Y}(\mathbf{s_Y}) = \begin{bmatrix} s_{Y1} & s_{Y2} \\ -s_{Y2}^* & s_{Y1}^* \end{bmatrix}$, since $\mathbf{X} = \mathbf{Y}(\mathbf{s_X}) = \mathbf{Y}(s_{X1}, s_{X2})$. To illustrate the case where $\mu_X = 1$ and $\mu_Y = 2$, $\mathbf{X}(\mathbf{s_X}) = \begin{bmatrix} s_{X1} & s_{X1} \\ -s_{X1}^* & s_{X1}^* \end{bmatrix}$ (which presents the Alamouti code with only one variable) is also of similar form to $\mathbf{Y}$ since $\mathbf{X} = \mathbf{Y}(\mathbf{s_X}) = \mathbf{Y}(s_{X1})$.

By this notation, when we state that the matrix $\mathbf{C}$ in (5.7) is of similar form to the matrix $\mathbf{B}$, for instance, we imply that $\mathbf{C}$ can be represented as $\mathbf{C} = k_C \mathbf{B}(\mathbf{s_C})$ where the number of complex variables μ_C in $\mathbf{C}$ is at most equal to the number of complex variables μ_B in $\mathbf{B}$, i.e., $\mu_C \leq \mu_B$.

5.3 Design Methods

In this section, we provide two new methods to construct square CO STBCs. In each case, we use sub-matrices of order n to build CO STBCs of order $N = 4n$. Our methods generalize the Williamson and Wallis-Whiteman arrays, which were originally used to build real orthogonal designs [Geramita and Seberry, 1979] (pp. 121 and 99, respectively).

PROPOSITION 5.3.1 *If the sub-matrices* $\mathbf{A}$, $\mathbf{B}$, $\mathbf{C}$ *and* $\mathbf{D}$ *of order n satisfy the following necessary conditions:*

1 $\mathbf{A}$, $\mathbf{B}$, $\mathbf{C}$ *and* $\mathbf{D}$ *are orthogonal themselves and*

$$\mathbf{A}^H\mathbf{A} + \mathbf{B}^H\mathbf{B} + \mathbf{C}^H\mathbf{C} + \mathbf{D}^H\mathbf{D} = \sum_{i\in\mathcal{I}_U} l_i|s_i|^2\mathbf{I}_n \qquad (5.8)$$

where l_i are definitely positive, real coefficients, and the complex variables s_i may be in U_1, U_2, U_3 or U_4, which are defined in (5.4);

2 The matrices $\mathcal{O}' = \begin{bmatrix} \mathbf{A} & \mathbf{B} \\ -\mathbf{B} & \mathbf{A} \end{bmatrix}$ *and* $\mathcal{O}'' = \begin{bmatrix} \mathbf{A} & \bar{\mathbf{B}} \\ -\bar{\mathbf{B}} & \mathbf{A} \end{bmatrix}$ *are square Complex Orthogonal Designs (CODs) of order $2n$;*

3 $\mathbf{B_s}^H\mathbf{B_{s'}}$ *and* $\mathbf{B_s}^T\mathbf{B_{s'}}$ *are symmetric for any possible pair of vectors $\mathbf{s}$ and $\mathbf{s'}$ of complex variables, where $\mathbf{B_s}$ and $\mathbf{B_{s'}}$ are shorthand for $\mathbf{B(s)}$ and $\mathbf{B(s')}$, respectively;*

4 *C and D are each of similar form to B, which implies that* $\mu_C \leq \mu_B$, $\mu_D \leq \mu_B$ *and*

$$\begin{cases} \mathbf{C} = k_C \mathbf{B}(\mathbf{s_C}) \\ \mathbf{D} = k_D \mathbf{B}(\mathbf{s_D}) \end{cases} \tag{5.9}$$

where k_C *and* k_D *are arbitrary (positive or negative), real coefficients,* $\mathbf{s_C}$ *is a vector containing variables* $s_{C1}, s_{C2}, \ldots, s_{C\mu_C}$, *and similarly for* $\mathbf{s_D}$;

then

$$\mathcal{O} = \begin{bmatrix} \mathbf{A} & \mathbf{B} & \mathbf{C} & \mathbf{D} \\ -\mathbf{B} & \mathbf{A} & \bar{\mathbf{D}} & -\bar{\mathbf{C}} \\ -\mathbf{C} & -\bar{\mathbf{D}} & \mathbf{A} & \bar{\mathbf{B}} \\ -\mathbf{D} & \bar{\mathbf{C}} & -\bar{\mathbf{B}} & \mathbf{A} \end{bmatrix} \tag{5.10}$$

is a COD of order $N = 4n$. *If all coefficients* $l_i = 1$ *for* $i \in \mathcal{I}_U$, *then* $\mathcal{O}$ *is called square CO STBC without Linear Processing (LP) (or just square CO STBC for short). Otherwise,* $\mathcal{O}$ *is considered as a square CO STBC with LP. In order for* $\mathcal{O}$ *to achieve the maximum number of variables (or maximum rates), we must have* $\mu_A + \mu_B + \mu_C + \mu_D = \mu(N)$.

Proof. From (5.10), we have the following equation:

$$
\begin{aligned}
\mathbf{M} &= \mathcal{O}^H \mathcal{O} \\
&= \begin{bmatrix}
\mathbf{A}^H\mathbf{A} + \mathbf{B}^H\mathbf{B} + \mathbf{C}^H\mathbf{C} + \mathbf{D}^H\mathbf{D} & \mathbf{A}^H\mathbf{B} - \mathbf{B}^H\mathbf{A} + \mathbf{C}^H\bar{\mathbf{D}} - \mathbf{D}^H\bar{\mathbf{C}} \\
 & \mathbf{A}^H\mathbf{A} + \mathbf{B}^H\mathbf{B} + \bar{\mathbf{C}}^H\bar{\mathbf{C}} + \bar{\mathbf{D}}^H\bar{\mathbf{D}} \\
 & \\
\mathcal{L} & \\
 & \\
\end{bmatrix}
\end{aligned}
$$

$$
\begin{bmatrix}
\mathbf{A}^H\mathbf{C} - \mathbf{C}^H\mathbf{A} - \mathbf{B}^H\bar{\mathbf{D}} + \mathbf{D}^H\bar{\mathbf{B}} & \mathbf{A}^H\mathbf{D} - \mathbf{D}^H\mathbf{A} + \mathbf{B}^H\bar{\mathbf{C}} - \mathbf{C}^H\bar{\mathbf{B}} \\
\mathbf{B}^H\mathbf{C} - \bar{\mathbf{C}}^H\bar{\mathbf{B}} + \mathbf{A}^H\bar{\mathbf{D}} - \bar{\mathbf{D}}^H\mathbf{A} & \mathbf{B}^H\mathbf{D} - \bar{\mathbf{D}}^H\bar{\mathbf{B}} - \mathbf{A}^H\bar{\mathbf{C}} + \bar{\mathbf{C}}^H\mathbf{A} \\
\mathbf{A}^H\mathbf{A} + \bar{\mathbf{B}}^H\bar{\mathbf{B}} + \mathbf{C}^H\mathbf{C} + \bar{\mathbf{D}}^H\bar{\mathbf{D}} & \mathbf{C}^H\mathbf{D} - \bar{\mathbf{D}}^H\bar{\mathbf{C}} + \mathbf{A}^H\bar{\mathbf{B}} - \bar{\mathbf{B}}^H\mathbf{A} \\
 & \mathbf{A}^H\mathbf{A} + \bar{\mathbf{B}}^H\bar{\mathbf{B}} + \bar{\mathbf{C}}^H\bar{\mathbf{C}} + \mathbf{D}^H\mathbf{D}
\end{bmatrix}
$$

$\mathcal{L}$ in the matrix $\mathbf{M}$ denotes the lower triangular part under the main diagonal whose elements are the Hermitian transposes of the corresponding elements in the upper triangular part. For instance, we have the element $\mathcal{L}(2,1) = \mathbf{B}^H\mathbf{A} - \mathbf{A}^H\mathbf{B} + \bar{\mathbf{D}}^H\mathbf{C} - \bar{\mathbf{C}}^H\mathbf{D}$.

First, we prove the following equalities:

$$\bar{\mathbf{B}}^H \bar{\mathbf{B}} = \mathbf{B}^H \mathbf{B} \tag{5.11}$$

$$\bar{\mathbf{C}}^H \bar{\mathbf{C}} = \mathbf{C}^H \mathbf{C} \tag{5.12}$$

$$\bar{\mathbf{D}}^H \bar{\mathbf{D}} = \mathbf{D}^H \mathbf{D} \tag{5.13}$$

Since **B** is orthogonal, we have

$$\mathbf{B}^H \mathbf{B} = \mathbf{B}\mathbf{B}^H = \sum_{i \in \mathcal{I}_{U_2}} l_i |s_i|^2 \mathbf{I}_n$$

which implies that $\mathbf{B}^H \mathbf{B}$ is a real, diagonal matrix and therefore

$$\mathbf{B}^H \mathbf{B} = [(\mathbf{B}^H \mathbf{B})^T]^H \tag{5.14}$$

Using Eq. (5.14), it follows that

$$\begin{aligned} \mathbf{B}^H \mathbf{B} &= [\mathbf{B}^T \bar{\mathbf{B}}]^H \\ &= \bar{\mathbf{B}}^H (\mathbf{B}^T)^H \\ &= \bar{\mathbf{B}}^H \bar{\mathbf{B}} \end{aligned}$$

Therefore, (5.11) has been proved. The same arguments can be applied to prove (5.12) and (5.13). Hence, if **A**, **B**, **C** and **D** are orthogonal themselves and satisfy (5.8), then all elements (i.e. sub-matrices) on the main diagonal of the matrix $\mathbf{M} = \mathcal{O}^H \mathcal{O}$ are equal to

$$\mathbf{A}^H \mathbf{A} + \mathbf{B}^H \mathbf{B} + \mathbf{C}^H \mathbf{C} + \mathbf{D}^H \mathbf{D} = \sum_{i \in \mathcal{I}_U} l_i |s_i|^2 \mathbf{I}_n$$

Second, we prove the following equalities:

$$\mathbf{A}^H \mathbf{B} - \mathbf{B}^H \mathbf{A} = \mathbf{O}_n \tag{5.15}$$

$$\mathbf{A}^H \mathbf{C} - \mathbf{C}^H \mathbf{A} = \mathbf{O}_n \tag{5.16}$$

$$\mathbf{A}^H \mathbf{D} - \mathbf{D}^H \mathbf{A} = \mathbf{O}_n \tag{5.17}$$

where $\mathbf{O}_n$ is a zero matrix of order n. Eq. (5.15) holds as $\mathcal{O}'$ is a COD. Additionally, because **C** and **D** are of similar form to **B** (see (5.9)), the equalities (5.16) and (5.17) are straightforwardly proved (multiplication with real coefficients k_C and k_D does not change the property (5.15)).

Third, we prove the following equalities:

$$\mathbf{B}^H \bar{\mathbf{C}} - \mathbf{C}^H \bar{\mathbf{B}} = \mathbf{O}_n \tag{5.18}$$

$$\mathbf{B}^H \bar{\mathbf{D}} - \mathbf{D}^H \bar{\mathbf{B}} = \mathbf{O}_n \tag{5.19}$$

$$\mathbf{C}^H \bar{\mathbf{D}} - \mathbf{D}^H \bar{\mathbf{C}} = \mathbf{O}_n \tag{5.20}$$

Since $\mathbf{B}_\mathbf{s}^T \mathbf{B}_{\mathbf{s}'}$ is symmetric for any pair of vectors $\mathbf{s}$ and $\mathbf{s}'$ of complex variables, it follows that $(\mathbf{B}_\mathbf{s}^T \mathbf{B}_{\mathbf{s}'})^H \equiv \mathbf{B}_{\mathbf{s}'}^H \mathbf{B}_\mathbf{s}^{H^T}$ is also symmetric. Using this symmetry, it follows that

$$\mathbf{B}_{\mathbf{s}'}^H \mathbf{B}_\mathbf{s}^{H^T} = \left[\mathbf{B}_{\mathbf{s}'}^H \mathbf{B}_\mathbf{s}^{H^T} \right]^T$$
$$\Leftrightarrow \mathbf{B}_{\mathbf{s}'}^H \bar{\mathbf{B}}_\mathbf{s} = \mathbf{B}_\mathbf{s}^H \mathbf{B}_{\mathbf{s}'}^{H^T}$$
$$\Leftrightarrow \mathbf{B}_{\mathbf{s}'}^H \bar{\mathbf{B}}_\mathbf{s} = \mathbf{B}_\mathbf{s}^H \bar{\mathbf{B}}_{\mathbf{s}'}$$

In other words, we have

$$\mathbf{B}_{\mathbf{s}'}^H \bar{\mathbf{B}}_\mathbf{s} - \mathbf{B}_\mathbf{s}^H \bar{\mathbf{B}}_{\mathbf{s}'} = \mathbf{O}_n \qquad (5.21)$$

for any pair of vectors $\mathbf{s}$ and $\mathbf{s}'$. Due to the fact that $\mathbf{C}$ and $\mathbf{D}$ are of similar form to $\mathbf{B}$, by replacing $\mathbf{B}_\mathbf{s}$ and $\mathbf{B}_{\mathbf{s}'}$ in (5.21) by $\mathbf{B}$, $\mathbf{C}$ or $\mathbf{D}$, the equalities (5.18), (5.19) and (5.20) are proved.

From (5.15)–(5.20), we see that the elements $\mathbf{M}(1,2)$, $\mathbf{M}(1,3)$ and $\mathbf{M}(1,4)$ of the matrix $\mathbf{M} = \mathcal{O}^H \mathcal{O}$ are zero matrices.

Fourth, we prove the following equalities:

$$\mathbf{B}^H \mathbf{C} - \bar{\mathbf{C}}^H \bar{\mathbf{B}} = \mathbf{O}_n \qquad (5.22)$$
$$\mathbf{B}^H \mathbf{D} - \bar{\mathbf{D}}^H \bar{\mathbf{B}} = \mathbf{O}_n \qquad (5.23)$$
$$\mathbf{C}^H \mathbf{D} - \bar{\mathbf{D}}^H \bar{\mathbf{C}} = \mathbf{O}_n \qquad (5.24)$$

Due to $\mathbf{B}_\mathbf{s}^H \mathbf{B}_{\mathbf{s}'}$ being symmetric, the following equalities hold for any pair of vectors $\mathbf{s}$ and $\mathbf{s}'$:

$$\mathbf{B}_\mathbf{s}^H \mathbf{B}_{\mathbf{s}'} = \left[\mathbf{B}_\mathbf{s}^H \mathbf{B}_{\mathbf{s}'} \right]^T$$
$$\Leftrightarrow \mathbf{B}_\mathbf{s}^H \mathbf{B}_{\mathbf{s}'} = \mathbf{B}_{\mathbf{s}'}^T \bar{\mathbf{B}}_\mathbf{s}$$
$$\Leftrightarrow \mathbf{B}_\mathbf{s}^H \mathbf{B}_{\mathbf{s}'} = \bar{\mathbf{B}}_{\mathbf{s}'}^H \bar{\mathbf{B}}_\mathbf{s}$$
$$\Leftrightarrow \mathbf{B}_\mathbf{s}^H \mathbf{B}_{\mathbf{s}'} - \bar{\mathbf{B}}_{\mathbf{s}'}^H \bar{\mathbf{B}}_\mathbf{s} = \mathbf{O}_n \qquad (5.25)$$

Due to $\mathbf{C}$ and $\mathbf{D}$ being of similar form to $\mathbf{B}$, by replacing $\mathbf{B}_\mathbf{s}$ and $\mathbf{B}_{\mathbf{s}'}$ in (5.25) by $\mathbf{B}$, $\mathbf{C}$ or $\mathbf{D}$, the equalities (5.22)–(5.24) are proved.

Finally, we prove that

$$\mathbf{A}^H \bar{\mathbf{B}} - \bar{\mathbf{B}}^H \mathbf{A} = \mathbf{O}_n \qquad (5.26)$$
$$\mathbf{A}^H \bar{\mathbf{C}} - \bar{\mathbf{C}}^H \mathbf{A} = \mathbf{O}_n \qquad (5.27)$$
$$\mathbf{A}^H \bar{\mathbf{D}} - \bar{\mathbf{D}}^H \mathbf{A} = \mathbf{O}_n \qquad (5.28)$$

Eq. (5.26) holds since $\mathcal{O}''$ is a COD. Because $\mathbf{C}$ and $\mathbf{D}$ are of similar form to $\mathbf{B}$, by replacing $\mathbf{B}$ in (5.26) by $\mathbf{C}$ or $\mathbf{D}$, the equalities (5.27) and (5.28) are proved.

From (5.22)–(5.24) and (5.26)–(5.28), it follows that the elements $\mathbf{M}(2,3)$ = $\mathbf{M}(2,4)$ = $\mathbf{M}(3,4)$ = $\mathbf{O}_n$. Since the lower triangular part $\mathcal{L}$ is the Hermitian transpose of the upper part, all elements in $\mathcal{L}$ are also zero matrices. Hence, $\mathbf{M}$ can be presented as

$$\mathbf{M} = \sum_{i \in \mathcal{I}_U} l_i |s_i|^2 diag(\mathbf{I}_n, \mathbf{I}_n, \mathbf{I}_n, \mathbf{I}_n) = \sum_{i \in \mathcal{I}_U} l_i |s_i|^2 \mathbf{I}_N$$

where $diag$ denotes a diagonal matrix. In other words, the matrix $\mathcal{O}$ in (5.10) is a square COD (also CO STBC) of order $N=4n$ with $(\mu_A + \mu_B + \mu_C + \mu_D)$ variables. Note that, if $\mathcal{O}$ comprises the maximum number of variables, i.e., Eq. (5.6) is satisfied, then $\mathcal{O}$ is a square, maximum rate CO STBC of order $4n$. The proposition has been proved. ∎

COROLLARY 5.3.2 *If the sub-matrices* **A**, **B**, **C** *and* **D** *of order n satisfy:*

1 The conditions (1), (2), (3) mentioned in the proposition 5.3.1;

2 **C** *and* **D** *are of similar form to* $\bar{\mathbf{B}}$ *and* **B**, *respectively,* **B** *and* $\bar{\mathbf{B}}$ *respectively, or* $\bar{\mathbf{B}}$ *and* $\bar{\mathbf{B}}$, *respectively, which implies that* $\mu_C \leq \mu_B$, $\mu_D \leq \mu_B$ *and* **C**, **D** *can be presented as one of the following forms:*

$$\begin{cases} \mathbf{C} = k_C \mathbf{B}(\mathbf{s}_\mathbf{C}^*) \\ \mathbf{D} = k_D \mathbf{B}(\mathbf{s}_\mathbf{D}) \end{cases}$$

$$\begin{cases} \mathbf{C} = k_C \mathbf{B}(\mathbf{s}_\mathbf{C}) \\ \mathbf{D} = k_D \mathbf{B}(\mathbf{s}_\mathbf{D}^*) \end{cases}$$

$$\begin{cases} \mathbf{C} = k_C \mathbf{B}(\mathbf{s}_\mathbf{C}^*) \\ \mathbf{D} = k_D \mathbf{B}(\mathbf{s}_\mathbf{D}^*) \end{cases}$$

where k_C *and* k_D *are arbitrary (positive or negative), real coefficients;*

then $\mathcal{O}$ *in (5.10) is a CO STBC of order* $N=4n$.

Proof. We prove the corollary for the case that **C** and **D** are of similar form to $\bar{\mathbf{B}}$ and **B**, respectively. Similar arguments can be applied to the two other cases. We must show that Eqs. (5.11)–(5.13), (5.15)–(5.17), (5.18)–(5.20), (5.22)–(5.24), and (5.26)–(5.28) hold under the conditions of the Corollary.

For ease of exposition, we denote

$$\mathbf{C} = k_C \bar{\mathbf{B}}_1 \qquad \mathbf{D} = k_D \mathbf{B}_2$$

Eqs. (5.11)–(5.13) hold due to the first condition of Proposition 5.3.1. The proof is not affected by the new conditions of this corollary.

To prove Eqs. (5.15)–(5.17), we use the second condition of Proposition 5.3.1. Eq. (5.15) holds because $\mathcal{O}'$ is a COD. Eq. (5.16) holds because

$$\mathbf{A}^H\mathbf{C} - \mathbf{C}^H\mathbf{A} \;=\; k_C(\mathbf{A}^H\bar{\mathbf{B}}_1 - \bar{\mathbf{B}}_1^H\mathbf{A}) = \mathbf{O}_n \qquad (5.29)$$

with the last equality in (5.29) holding because $\mathcal{O}''$ is a COD. Similar arguments are applied to prove (5.17).

The equality (5.20) holds because

$$\begin{aligned}
\mathbf{C}^H\bar{\mathbf{D}} - \mathbf{D}^H\bar{\mathbf{C}} &\;=\; k_C k_D(\mathbf{B}_1^T\mathbf{B}_2^{H^T} - \mathbf{B}_2^H\mathbf{B}_1) \\
&\;=\; k_C k_D[(\mathbf{B}_2^H\mathbf{B}_1)^T - \mathbf{B}_2^H\mathbf{B}_1] \\
&\;=\; \mathbf{O}_n
\end{aligned}$$

with the last equality due to the fact that $\mathbf{B}_2^H\mathbf{B}_1$ is symmetric (the 3rd condition in Proposition 5.3.1). Similar arguments are applied to prove (5.18) and (5.19).

The equality (5.22) holds because

$$\mathbf{B}^H\mathbf{C} - \bar{\mathbf{C}}^H\bar{\mathbf{B}} \;=\; k_C\big\{(\mathbf{B}_1^T\mathbf{B})^H - [(\mathbf{B}_1^T\mathbf{B})^H]^T\big\} = \mathbf{O}_n$$

with the last equality due to $\mathbf{B}_1^T\mathbf{B}$ being symmetric (the 3rd condition in Proposition 5.3.1). Similar arguments are applied to prove (5.23) and (5.24).

Eq. (5.26) obviously holds since $\mathcal{O}''$ is a COD. The equality (5.27) holds because

$$\mathbf{A}^H\bar{\mathbf{C}} - \bar{\mathbf{C}}^H\mathbf{A} \;=\; k_C\big[\mathbf{A}^H\mathbf{B}_1 - \mathbf{B}_1^H\mathbf{A}\big] = \mathbf{O}_n$$

with the last equality due to $\mathcal{O}'$ being a COD (the 2rd condition in Proposition 5.3.1). Similar arguments are applied to prove (5.28). The corollary has been proved.   ∎

We summarize the Proposition 5.3.1 and the Corollary 5.3.2 into the following theorem which is a modification of the Williamson array [Geramita and Seberry, 1979] (pp. 121) in order to apply to complex matrices:

THEOREM 5.3.3 *If the sub-matrices* $\mathbf{A}$, $\mathbf{B}$, $\mathbf{C}$ *and* $\mathbf{D}$ *of order* n *satisfy the following necessary conditions:*

1 $\mathbf{A}$, $\mathbf{B}$, $\mathbf{C}$, *and* $\mathbf{D}$ *are orthogonal themselves and*

$$\mathbf{A}^H\mathbf{A} + \mathbf{B}^H\mathbf{B} + \mathbf{C}^H\mathbf{C} + \mathbf{D}^H\mathbf{D} = \sum_{i \in \mathcal{I}_U} l_i|s_i|^2\mathbf{I}_n$$

where l_i are definitely positive, real coefficients, and the complex variables s_i may be in U_1, U_2, U_3 or U_4 which are defined in (5.4).

2 *The matrices* $\mathcal{O}' = \begin{bmatrix} \mathbf{A} & \mathbf{B} \\ -\mathbf{B} & \mathbf{A} \end{bmatrix}$ *and* $\mathcal{O}'' = \begin{bmatrix} \mathbf{A} & \bar{\mathbf{B}} \\ -\bar{\mathbf{B}} & \mathbf{A} \end{bmatrix}$ *are square Complex Orthogonal Designs (COD) of order $2n$.*

3 $\mathbf{B}_\mathbf{s}^H \mathbf{B}_{\mathbf{s}'}$ *and* $\mathbf{B}_\mathbf{s}^T \mathbf{B}_{\mathbf{s}'}$ *are symmetric for any possible pair of vectors* $\mathbf{s}$ *and* $\mathbf{s}'$ *of complex variables, where* $\mathbf{B}_\mathbf{s}$ *and* $\mathbf{B}_{\mathbf{s}'}$ *are shorthand for* $\mathbf{B}(\mathbf{s})$ *and* $\mathbf{B}(\mathbf{s}')$, *respectively.*

4 $\mathbf{C}$ *and* $\mathbf{D}$ *are of similar form to* $\mathbf{B}$ *and* $\mathbf{B}$, *respectively,* $\bar{\mathbf{B}}$ *and* $\mathbf{B}$, *respectively,* $\mathbf{B}$ *and* $\bar{\mathbf{B}}$ *respectively, or* $\bar{\mathbf{B}}$ *and* $\bar{\mathbf{B}}$, *respectively, i.e.,* $\mathbf{C}$ *and* $\mathbf{D}$ *can be presented as one of the following forms:*

$$\begin{cases} \mathbf{C} = k_C \mathbf{B}(\mathbf{s_C}) \\ \mathbf{D} = k_D \mathbf{B}(\mathbf{s_D}) \end{cases}$$

$$\begin{cases} \mathbf{C} = k_C \mathbf{B}(\mathbf{s_C^*}) \\ \mathbf{D} = k_D \mathbf{B}(\mathbf{s_D}) \end{cases}$$

$$\begin{cases} \mathbf{C} = k_C \mathbf{B}(\mathbf{s_C}) \\ \mathbf{D} = k_D \mathbf{B}(\mathbf{s_D^*}) \end{cases}$$

$$\begin{cases} \mathbf{C} = k_C \mathbf{B}(\mathbf{s_C^*}) \\ \mathbf{D} = k_D \mathbf{B}(\mathbf{s_D^*}) \end{cases}$$

where k_C and k_D are arbitrary (positive or negative), real coefficients, and $\mu_C \leq \mu_B$, $\mu_D \leq \mu_B$

then

$$\mathcal{O} = \begin{bmatrix} \mathbf{A} & \mathbf{B} & \mathbf{C} & \mathbf{D} \\ -\mathbf{B} & \mathbf{A} & \bar{\mathbf{D}} & -\bar{\mathbf{C}} \\ -\mathbf{C} & -\bar{\mathbf{D}} & \mathbf{A} & \bar{\mathbf{B}} \\ -\mathbf{D} & \bar{\mathbf{C}} & -\bar{\mathbf{B}} & \mathbf{A} \end{bmatrix} \qquad (5.30)$$

is a CO STBC of order $N=4n$. If all coefficients $l_i = 1$, for $i \in \mathcal{I}_U$, then $\mathcal{O}$ is called square CO STBC without Linear Processing (LP) (or just square CO STBC for short). Otherwise, $\mathcal{O}$ is considered as a square CO STBC with LP. If $(\mu_A + \mu_B + \mu_C + \mu_D) = \mu(N)$, then $\mathcal{O}$ is a square, maximum rate CO STBC of order $4n$.

Similarly, we derived the following theorem, which is a variation of the Wallis-Whiteman array [Geramita and Seberry, 1979] (pp. 99), modified to apply to complex matrices:

THEOREM 5.3.4 *If the sub-matrices* $\mathbf{A}$, $\mathbf{B}$, $\mathbf{C}$ *and* $\mathbf{D}$ *of order* n *satisfy the following necessary conditions:*

1 $\mathbf{A}$, $\mathbf{B}$, $\mathbf{C}$ *and* $\mathbf{D}$ *are orthogonal themselves and*

$$\mathbf{A}^H\mathbf{A} + \mathbf{B}^H\mathbf{B} + \mathbf{C}^H\mathbf{C} + \mathbf{D}^H\mathbf{D} = \sum_{i \in \mathcal{I}_U} l_i|s_i|^2\mathbf{I}_n$$

where l_i *are definitely positive, real coefficients, and the complex variables* s_i *may be in* U_1, U_2, U_3 *or* U_4, *which are defined in (5.4).*

2 The matrices $\mathcal{O}' = \begin{bmatrix} \mathbf{C} & \mathbf{A} \\ -\mathbf{A} & \mathbf{C} \end{bmatrix}$ *and* $\mathcal{O}'' = \begin{bmatrix} \mathbf{C} & \bar{\mathbf{A}} \\ -\bar{\mathbf{A}} & \mathbf{C} \end{bmatrix}$ *are square Complex Orthogonal Designs (CODs) of order* $2n$.

3 $\mathbf{A}_\mathbf{s}^H\mathbf{A}_{\mathbf{s}'}$ *and* $\mathbf{A}_\mathbf{s}^T\mathbf{A}_{\mathbf{s}'}$ *are symmetric for any possible pair of vectors* $\mathbf{s}$ *and* $\mathbf{s}'$ *of complex variables, where* $\mathbf{A}_\mathbf{s}$ *and* $\mathbf{A}_{\mathbf{s}'}$ *are shorthand for* $\mathbf{A}(\mathbf{s})$ *and* $\mathbf{A}(\mathbf{s}')$, *respectively.*

4 $\mathbf{B}$ *and* $\mathbf{D}$ *are of similar form to* $\mathbf{A}$ *and* $\mathbf{A}$, *respectively,* $\bar{\mathbf{A}}$ *and* $\mathbf{A}$, *respectively,* $\mathbf{A}$ *and* $\bar{\mathbf{A}}$ *respectively, or* $\bar{\mathbf{A}}$ *and* $\bar{\mathbf{A}}$, *respectively, i.e.,* $\mathbf{B}$ *and* $\mathbf{D}$ *can be presented as one of the following forms:*

$$\begin{cases} \mathbf{B} = k_B\mathbf{A}(\mathbf{s_B}) \\ \mathbf{D} = k_D\mathbf{A}(\mathbf{s_D}) \end{cases}$$

$$\begin{cases} \mathbf{B} = k_B\mathbf{A}(\mathbf{s_B^*}) \\ \mathbf{D} = k_D\mathbf{A}(\mathbf{s_D}) \end{cases}$$

$$\begin{cases} \mathbf{B} = k_B\mathbf{A}(\mathbf{s_B}) \\ \mathbf{D} = k_D\mathbf{A}(\mathbf{s_D^*}) \end{cases}$$

$$\begin{cases} \mathbf{B} = k_B\mathbf{A}(\mathbf{s_B^*}) \\ \mathbf{D} = k_D\mathbf{A}(\mathbf{s_D^*}) \end{cases}$$

where k_B *and* k_D *are arbitrary (positive or negative), real coefficients, and* $\mu_B \leq \mu_A$, $\mu_D \leq \mu_A$

then

$$\mathcal{O} = \begin{bmatrix} \mathbf{A} & \mathbf{B} & \mathbf{C} & \mathbf{D} \\ -\bar{\mathbf{B}} & \bar{\mathbf{A}} & -\mathbf{D} & \mathbf{C} \\ -\mathbf{C} & \bar{\mathbf{D}} & \mathbf{A} & -\bar{\mathbf{B}} \\ -\bar{\mathbf{D}} & -\mathbf{C} & \mathbf{B} & \bar{\mathbf{A}} \end{bmatrix} \qquad (5.31)$$

is a CO STBC of order $N=4n$. If all coefficients $l_i= 1$ for $i \in \mathcal{I}_U$, then $\mathcal{O}$ is called square CO STBC without Linear Processing (LP) (or just square CO STBC for short). Otherwise, $\mathcal{O}$ is considered as a square CO STBC with LP. If $(\mu_A + \mu_B + \mu_C + \mu_D)=\mu(N)$, then $\mathcal{O}$ is a square, maximum rate CO STBC of order $4n$.

Proof. The proof of Theorem 5.3.4 is similar to the proofs of Proposition 5.3.1 and Corollary 5.3.2.   ∎

5.4   Examples of Maximum Rate, Square, Order-8 CO STBCs with No Zero Entries

In order to construct 8×8 CO STBCs of the maximum rate using the proposed methods in Theorems 5.3.3 and 5.3.4, the main task is to find two 2×2 sub-matrices which satisfy certain properties. This is easier than finding eight 8×8 weighting matrices simultaneously as in the AOD approach [Yuen et al., 2004].

Using Theorem 5.3.3 and Theorem 5.3.4, we construct here some *square* CO STBCs of order $N=8$ (with or without LP) with the maximum number of variables $\mu(8)=4$. The sub-matrices $\mathbf{A}$, $\mathbf{B}$, $\mathbf{C}$, $\mathbf{D}$ are of order $n = 2$ and each sub-matrix comprises one variable.   From Theorem 5.3.3 (correspondingly, Theorem 5.3.4), it is clear that the most *crucial* task for constructing square CO STBCs of order $4n$ in our proposed methods is to find two matrices $\mathbf{A}$ and $\mathbf{B}$ ($\mathbf{A}$ and $\mathbf{C}$) satisfying the properties (2) and (3) in Theorem 5.3.3 (Theorem 5.3.4). We realize that various matrices $\mathbf{A}$, $\mathbf{B}$, $\mathbf{C}$, and $\mathbf{D}$ can satisfy those conditions, and derive here some of those cases for illustration.

EXAMPLE 5.4.1 *The following sub-matrices satisfy Theorem 5.3.3:*

$$\mathbf{A} = k_1 \begin{bmatrix} s_1 & s_1 \\ -s_1^* & s_1^* \end{bmatrix} ; \mathbf{B} = k_2 \begin{bmatrix} -s_2^* & s_2^* \\ s_2 & s_2 \end{bmatrix} ;$$

$$\mathbf{C} = k_3 \begin{bmatrix} -s_3^* & s_3^* \\ s_3 & s_3 \end{bmatrix} ; \mathbf{D} = k_4 \begin{bmatrix} -s_4^* & s_4^* \\ s_4 & s_4 \end{bmatrix}$$

for any real coefficients k_i, $(i = 1 \ldots 4)$.

In this example, $\mathbf{A}$ is a variation of the Alamouti code with only one variable, while $\mathbf{C}$ and $\mathbf{D}$ are each of similar form to $\mathbf{B}$. Then, $\mathcal{O}$ in (5.30) satisfies $\mathcal{O}^H \mathcal{O} = 2 \sum_{i=1}^4 k_i^2 |s_i|^2 \mathbf{I}_8$ and, consequently, $\mathcal{O}$ is a square, maximum rate CO STBC of order 8 (with or without LP depending on $k_i s$). If $k_i=1$, for $i = 1, \ldots, 4$, from (5.30), we have the following code:

$$\mathcal{O} = \begin{bmatrix} s_1 & s_1 & -s_2^* & s_2^* & -s_3^* & s_3^* & -s_4^* & s_4^* \\ -s_1^* & s_1^* & s_2 & s_2 & s_3 & s_3 & s_4 & s_4 \\ s_2^* & -s_2^* & s_1 & s_1 & -s_4 & s_4 & s_3 & -s_3 \\ -s_2 & -s_2 & -s_1^* & s_1^* & s_4^* & s_4^* & -s_3^* & -s_3^* \\ s_3^* & -s_3^* & s_4 & -s_4 & s_1 & s_1 & -s_2 & s_2 \\ -s_3 & -s_3 & -s_4^* & -s_4^* & -s_1^* & s_1^* & s_2^* & s_2^* \\ s_4^* & -s_4^* & -s_3 & s_3 & s_2 & -s_2 & s_1 & s_1 \\ -s_4 & -s_4 & s_3^* & s_3^* & -s_2^* & -s_2^* & -s_1^* & s_1^* \end{bmatrix} \quad (5.32)$$

Examples with various other structures are given below.

EXAMPLE 5.4.2 *This example illustrates the case in Theorem 5.3.3 where* **C** *and* **D** *are each of similar form to* $\bar{\mathbf{B}}$:

$$\mathbf{A} = k_1 \begin{bmatrix} s_1 & -s_1 \\ s_1^* & s_1^* \end{bmatrix}; \mathbf{B} = k_2 \begin{bmatrix} s_2^* & s_2^* \\ s_2 & -s_2 \end{bmatrix};$$

$$\mathbf{C} = k_3 \begin{bmatrix} s_3 & s_3 \\ s_3^* & -s_3^* \end{bmatrix}; \mathbf{D} = k_4 \begin{bmatrix} s_4 & s_4 \\ s_4^* & -s_4^* \end{bmatrix}$$

If $k_1 = k_2 = 1$, *and* $k_3 = k_4 = -1$, *from (5.30), we have the following code:*

$$\mathcal{O} = \begin{bmatrix} s_1 & -s_1 & s_2^* & s_2^* & -s_3 & -s_3 & -s_4 & -s_4 \\ s_1^* & s_1^* & s_2 & -s_2 & -s_3^* & s_3^* & -s_4^* & s_4^* \\ -s_2^* & -s_2^* & s_1 & -s_1 & -s_4^* & -s_4^* & s_3^* & s_3^* \\ -s_2 & s_2 & s_1^* & s_1^* & -s_4 & s_4 & s_3 & -s_3 \\ s_3 & s_3 & s_4^* & s_4^* & s_1 & -s_1 & s_2 & s_2 \\ s_3^* & -s_3^* & s_4 & -s_4 & s_1^* & s_1^* & s_2^* & -s_2^* \\ s_4 & s_4 & -s_3^* & -s_3^* & -s_2 & -s_2 & s_1 & -s_1 \\ s_4^* & -s_4^* & -s_3 & s_3 & -s_2^* & s_2^* & s_1^* & s_1^* \end{bmatrix} \quad (5.33)$$

EXAMPLE 5.4.3 *This example using Theorem 5.3.3 shows that the CO STBC* **G**8 *in (5.1) can be (indirectly) derived from our proposed methods. Let:*

$$\mathbf{A} = k_1 \begin{bmatrix} s_1^* & s_1^* \\ s_1 & -s_1 \end{bmatrix}; \mathbf{B} = k_2 \begin{bmatrix} s_2 & -s_2 \\ s_2^* & s_2^* \end{bmatrix};$$

$$\mathbf{C} = k_3 \begin{bmatrix} s_3 & -s_3 \\ s_3^* & s_3^* \end{bmatrix}; \mathbf{D} = k_4 \begin{bmatrix} s_4 & -s_4 \\ s_4^* & s_4^* \end{bmatrix}$$

If $k_i = 1$ for $i = 1, \ldots, 4$, from (5.30), we have the following code:

$$
\mathcal{O} = \begin{bmatrix}
s_1^* & s_1^* & s_2 & -s_2 & s_3 & -s_3 & s_4 & -s_4 \\
s_1 & -s_1 & s_2^* & s_2^* & s_3^* & s_3^* & s_4^* & s_4^* \\
-s_2 & s_2 & s_1^* & s_1^* & s_4^* & -s_4^* & -s_3^* & s_3^* \\
-s_2^* & -s_2^* & s_1 & -s_1 & s_4 & s_4 & -s_3 & -s_3 \\
-s_3 & s_3 & -s_4^* & s_4^* & s_1^* & s_1^* & s_2^* & -s_2^* \\
-s_3^* & -s_3^* & -s_4 & -s_4 & s_1 & -s_1 & s_2 & s_2 \\
-s_4 & s_4 & s_3^* & -s_3^* & -s_2^* & s_2^* & s_1^* & s_1^* \\
-s_4^* & -s_4^* & s_3 & s_3 & -s_2 & -s_2 & s_1 & -s_1
\end{bmatrix}
\tag{5.34}
$$

We note that the CO STBC $\mathbf{G}8$ in (5.1) can be derived from our CO STBC in (5.34) by multiplying every even row in (5.34) with j. However, $\mathbf{G}8$ in (5.1) itself does not follow our proposed structure as the sub-matrices $\mathbf{A}$ and $\mathbf{B}$ in $\mathbf{G}8$ do not satisfy the second condition in Theorem 5.3.3.

EXAMPLE 5.4.4 *This example illustrates the case in Theorem 5.3.4 where $\mathbf{B}$ and $\mathbf{D}$ are each of similar form to $\mathbf{A}$:*

$$
\mathbf{A} = k_1 \begin{bmatrix} s_1^* & s_1^* \\ s_1 & -s_1 \end{bmatrix} ; \mathbf{B} = k_2 \begin{bmatrix} s_2^* & s_2^* \\ s_2 & -s_2 \end{bmatrix} ;
$$

$$
\mathbf{C} = k_3 \begin{bmatrix} s_3 & -s_3 \\ s_3^* & s_3^* \end{bmatrix} ; \mathbf{D} = k_4 \begin{bmatrix} s_4^* & s_4^* \\ s_4 & -s_4 \end{bmatrix}
$$

If $k_i = 1$ for $i = 1, \ldots, 4$, from (5.31), we have the following code:

$$
\mathcal{O} = \begin{bmatrix}
s_1^* & s_1^* & s_2^* & s_2^* & s_3 & -s_3 & s_4^* & s_4^* \\
s_1 & -s_1 & s_2 & -s_2 & s_3^* & s_3^* & s_4 & -s_4 \\
-s_2 & -s_2 & s_1 & s_1 & -s_4^* & -s_4^* & s_3 & -s_3 \\
-s_2^* & s_2^* & s_1^* & -s_1^* & -s_4 & s_4 & s_3^* & s_3^* \\
-s_3 & s_3 & s_4 & s_4 & s_1^* & s_1^* & -s_2 & -s_2 \\
-s_3^* & -s_3^* & s_4^* & -s_4^* & s_1 & -s_1 & -s_2^* & s_2^* \\
-s_4 & -s_4 & -s_3 & s_3 & s_2^* & s_2^* & s_1 & s_1 \\
-s_4^* & s_4^* & -s_3^* & -s_3^* & s_2 & -s_2 & s_1^* & -s_1^*
\end{bmatrix}
\tag{5.35}
$$

All of the above codes are square, maximum rate CO STBCs of order $N=8$ with a full design, i.e., without any zeros for any complex signal constellation. The power is equally transmitted via each Tx antenna during every symbol time slot. For these reasons, the proposed CO STBCs are referred to as the *improved, square CO STBCs.*

5.5   Conclusions

By modifying the Williamson and the Wallis-Whiteman arrays to apply to complex matrices, we have proposed two new methods of constructing square, order-$4n$ CO STBCs from square, order-n CO STBCs which satisfy certain properties as described in Theorems 5.3.3 and 5.3.4. Although constructions of square, maximum rate CO STBCs, such as the Adams-Lax-Phillips construction, Jozefiak construction, and Wolfe construction, are well known [Liang, 2003], the codes resulting from these constructions have numerous zeros, since these construction methods always involve identity matrices. This disadvantage is overcome by our proposed constructions, where the sub-matrices are wisely selected to include as few zeros as possible or even no zeros. A smaller number of zeros in the sub-matrices results in a smaller number of zeros in the resultant CO STBC.

Our methods also partially overcome the difficulty in designing codes from AODs. For instance, in order to construct square, maximum rate codes of order 8, our methods mainly require us to find two 2×2 matrices satisfying our conditions, as opposed to requiring us to find eight 8×8 weighting matrices as in the AOD construction.

By applying Theorems 5.3.3 and 5.3.4, we have constructed various square, maximum rate, order-8 CO STBCs with no zeros. In our CO STBCs, the transmitted symbols equally disperse through Tx antennas with the consequence that the power transmitted via each Tx antenna is equal during every symbol time slot. Additionally, our methods may be used to design square CO STBCs of order 16 or 32 from square CO STBCs of order 4 or 8, respectively, provided that there exist sub-matrices satisfying the conditions of our theorems. The construction of square CO STBCs of higher orders, such as 16 or 32, requires further study.

Chapter 6

TRANSMITTER DIVERSITY ANTENNA SELECTION TECHNIQUES FOR MIMO SYSTEMS

6.1　Introduction

In wireless communication systems, the performance of downlink channels can be improved by transmission diversity techniques utilizing multiple transmitter antennas (Tx antennas) at base stations. A variety of transmission diversity techniques have been proposed in the literature so far, including beamforming, antenna switching, delay transmission etc., such as in [Blogh and Hanzo, 2002], [Electronics, 2002], [Katz and Ylitalo, 2000], [Lo and Tarokh, 1999].

The diversity combination of Space-Time Block Codes (STBCs) and a closed loop antenna selection technique (AST) assisted by a feedback loop to improve the performance of MIMO channels using STBCs with *coherent detection* has been intensively examined in the literature, such as [Chen et al., 2003a], [Chen et al., 2003c], [Gore and Paulraj, 2001], [Gore and Paulraj, 2002], [Katz et al., 2001], [Xiaofeng et al., 2001]. This combination provides a remarkable improvement in the error performance in both flat and frequency selective Rayleigh fading channels.

One simple and interesting transmitter diversity antenna selection technique (AST) was proposed by M. Katz et al. [Katz et al., 2001]. According to this technique, an M-antenna transmitter and one-antenna receiver are considered. The receiver measures M channel gains from M Tx antennas. Based on the measurement, the receiver informs the transmitter via a feedback loop about the N best channels ($N < M$). The authors in [Katz et al., 2001] named this technique as the N-out-of-M antenna selection technique (N-out-of-M AST). In this technique, the receiver uses $\lceil log_2\binom{M}{N}\rceil$ ($\lceil . \rceil$ is the ceiling function) feedback bits to inform the transmitter about the N best channels out of M channels. This AST is easily generalized for the case where the receiver has K receiver antennas (K Rx antennas).

In [Katz et al., 2001], the authors also mentioned a modified technique of the N-out-of-M AST, which was referred to as the *restricted* N-out-of-M AST. In the *restricted* N-out-of-M AST, the capacity limitation of the feedback loop was taken into account. Particularly, in this technique, the receiver uses only one feedback bit to inform the transmitter. In order to do that, M Tx antennas are divided into two subsets each of which contains N Tx antennas. Subsets may overlap one another. Let us consider the case where $M = 4$ and $N = 2$ as an example. Based on the total power received from antenna pairs (1,2) and (3,4), the receiver informs the transmitter about the antenna pair which should be selected. Obviously, the penalty of this method is performance degradation, compared to the N-out-of-M AST, since the Tx antenna pair from which the received total power is greater than the other is not necessary the pair of the two best Tx antennas.

Motivated by the N-out-of-M AST proposed in [Katz et al., 2001], we first propose a simple, improved, closed loop AST, which is referred to as the $(N + 1, N; K)$ AST/STBC scheme, to improve further the performance of wireless channels using STBCs with *coherent detection*. The notation $(N + 1, N; K)$ AST/STBC means that N Tx antennas are selected from $(N + 1)$ Tx antennas (based on the criterion mentioned later in this chapter) to transmit signals, while all K Rx antennas are used without selection. Our technique provides the same bit error performance as the N-out-of-$(N + 1)$ AST proposed in [Katz et al., 2001] if the receivers in both techniques have K Rx antennas. However, the $(N + 1, N; K)$ AST/STBC requires a shorter time to process the feedback information due to our proposed structure of the feedback information. Calculations and simulations show that our $(N + 1, N; K)$ AST/STBC scheme improves significantly the performance of channels using STBCs, especially when the Alamouti STBC is used.

As opposed to ASTs for channels utilizing STBCs with *coherent detection*, ASTs for channels utilizing Differential Space-Time Block Codes (DSTBCs) with *differential detection* have not been considered yet. The background knowledge on DSTBCs can be found in [Chen et al., 2003d], [Ganesan and Stoica, 2002a], [Ganesan and Stoica, 2002b], [Hochwald and Sweldens, 2000], [Hughes, 2000], [Tarokh and Jafarkhani, 2000], [Vucetic and Yuan, 2003].

In the rest of this chapter, we propose some ASTs which tend to maximize the signal-to-noise ratio (SNR) of channels using DSTBCs with *arbitrary* numbers of Tx and Rx antennas. We first propose the so-called *general* $(M, N; K)$ AST/DSTBC, where the transmitter selects N Tx antennas out of M Tx antennas $(M > N)$ to maximize the channel SNR. The antenna selection (at the transmitter) is based on the results of the comparison carried out (at the receiver) between the instantaneous powers of signals which are received during the initial transmission. The *general* $(M, N; K)$ AST/DSTBC improves significantly the performance of channels using DSTBCs. However, when M and N grow large, the number of feedback bits required to inform the transmitter also grows large. This drawback impedes the *general* $(M, N; K)$ AST/DSTBC from practical implementation if M and N are large.

The aforementioned drawback can be overcome by reducing the number of feedback bits. Based on this observation, we modify the *general* $(M, N; K)$ AST/DSTBC to derive the so-called *restricted* $(M, N; K)$ AST/DSTBC, which provides good bit error performance while using only one feedback bit for transmission diversity purpose.

Simulations show that STBCs and DSTBCs associated with the proposed ASTs provide much better bit error performance than those without using antenna selection. In the case of DSTBCs associated with our proposed ASTs, the improvement can be achieved with a limited number (typically, one or two) of training symbols and with a slightly modified receiver structure, compared to the case of conventional DSTBCs without our ASTs.

Although, the authors propose here the ASTs for a very general case, where the system may contain *arbitrary* numbers of Tx and Rx antennas, it is important to have in mind that it is *more practical* to have diversity antennas installed at the transmitter, e.g. at base stations in mobile communication systems, rather than at the hand-held, tiny receiver, such as mobile phones. It is well known that the installation of more than 2 Rx antennas in mobile phones is almost impractical due to the short life-time of batteries and due to the small size of the phones.

Consequently, by using the term "*antenna selection*" in this chapter, we mean *transmitter diversity* antenna selection, rather than *receiver diversity* antenna selection, i.e., all K Rx antennas are used *without* selection (although the generalization of the proposed ASTs for receiver diversity antenna selection is straightforward). It should be also noted that the terms "space-time block codes (STBCs)" and "differential space-time block codes (DSTBCs)" used throughout this chapter mean *complex, orthogonal* STBCs and *complex, orthogonal* DSTBCs, respectively.

The content of this chapter has been published in our following works [Tran and Wysocki, 2003], [Tran and Wysocki, 2004], [Tran et al., 2003], and [Tran et al., 2004b].

This chapter is organized as follows:

Section 6.2.1 introduces the theoretical basis of antenna selection in channels using STBCs with *coherent detection*. Section 6.2.2 includes derivation of the proposed $(N + 1, N; K)$ AST/STBC scheme for channels using STBCs with *coherent detection*. Simulation results for this improved AST are presented in Section 6.2.3.

Section 6.3.1 contains the review of the conventional DSTBCs mentioned in the literature and provides some remarks on the time-varying Rayleigh fading channels where DSTBCs can be practically used.

In Section 6.3.2, we mention some notations and assumptions used throughout this chapter. Section 6.3.3 starts with the discussion on the criterion of antenna selection in channels using STBCs and then analyzes our modifications to apply to channels using DSTBCs.

In Sections 6.3.4 and 6.3.5, we propose the *general* $(M, N; K)$ AST/DSTBC scheme and the *restricted* $(M, N; K)$ AST/DSTBC scheme, respectively.

In Section 6.3.6, we give some comments on the spatial diversity order of our proposed AST/DSTBC schemes. Simulation results are presented in Section 6.3.7 and the chapter is concluded by Section 6.4.

6.2   Improved Antenna Selection Technique for Wireless Channels Utilizing STBCs

6.2.1   Theoretical Basis of Antenna Selection in Wireless Channels Using STBCs with Coherent Detection

The core idea of antenna selection in channels using STBCs (with coherent detection), where transmission coefficients are assumed to be perfectly known at the receiver, has been intensively mentioned in the literature. Readers may refer to Section III A in [Gore and Paulraj, 2002] for instance. In this section, we derive a more explicit review of the antenna selection algorithm for channels using STBCs with the exact channel knowledge (ECK) at the receiver.

Let us consider a wireless system with N Tx antennas and K Rx antennas. The channel is assumed to be block flat fading. In other words, the transmission coefficients remain constant during multiple symbol time slots (STSs). Hence, the channel model is

$$\mathbf{Y} = \mathbf{HX} + \mathbf{N}$$

where $\mathbf{Y}_{(K \times T)}$, $\mathbf{X}_{(N \times T)}$, $\mathbf{H}_{(K \times N)}$ and $\mathbf{N}_{(K \times T)}$ are the matrices of received signals, transmitted signals, transmission coefficients and noises, respectively. Transmission coefficients and noises are assumed to be identically independently distributed (i.i.d.) complex Gaussian random variables with the distributions $\mathcal{CN}(0, 1)$ and $\mathcal{CN}(0, N_0)$, respectively. T is the number of STSs in each code block. Let E_S be the average energy per transmitted symbol and $\aleph$ be the number of transmitted symbols in each code block. Due to the orthogonality of STBCs, at the receiver, we have $\aleph$ independent decision metrics for $\aleph$ transmitted symbols $\{x_\eta\}$ $(\eta = 1, \ldots, \aleph)$ as follows [Tarokh et al., 1999a]:

$$\hat{x}_\eta = \sum_{i=1}^{K} \sum_{j=1}^{N} |h_{ij}|^2 x_\eta + \Xi_\eta \qquad (6.1)$$

where Ξ_η is a random variable with a zero mean and a variance of $N_0 \sum_{i=1}^{K} \sum_{j=1}^{N} |h_{ij}|^2$ for given h_{ij}s. From (6.1), we have the SNR of the ηth decision variable as

$$\gamma_\eta = \gamma_0 \sum_{i=1}^{K} \sum_{j=1}^{N} |h_{ij}|^2$$

where $\gamma_0 = \frac{E_S}{N_0}$ is the SNR of each transmitted symbol. Let $\xi_j \equiv \sum_{i=1}^{K} |h_{ij}|^2$ for $j = 1, \ldots, N$. We can rewrite the above equation as

$$\gamma_\eta = \gamma_0 \sum_{j=1}^{N} \xi_j \qquad (6.2)$$

We now consider a system including M Tx antennas and K Rx antennas and transmitting a square, order-N STBC where $N < M$. We select the N best Tx antennas among M Tx antennas to transmit signals. Clearly, the optimal *transmitter* antenna selection (in terms of maximizing SNR) is selecting N Tx antennas of which the transmission coefficients maximize the formula (6.2). Therefore, the N best Tx antennas which should be selected are corresponding to the N largest values among M values $\{\xi_1, \xi_2, \ldots, \xi_{M-1}, \xi_M\}$ to maximize SNR in (6.2). From the mathematical point of view, the algorithm selects the N out of M columns of the transmission coefficient matrix $\mathbf{H}$ with the highest Frobenius norms. This is the basis for selecting Tx antennas in channels using STBCs with ECK at the receiver [Gore and Paulraj, 2002]. An example of such ASTs is the one mentioned in [Katz et al., 2001], which was referred to as the N-out-of-M AST.

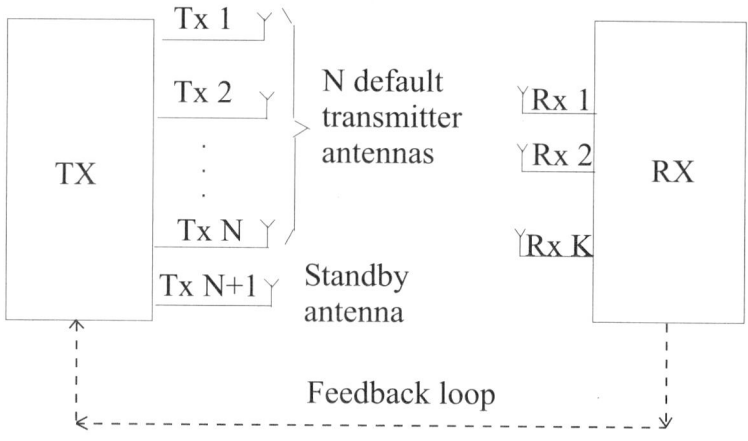

Figure 6.1.   The diagram of the $(N+1, N; K)$ AST/STBC scheme.

6.2.2   The $(N+1, N; K)$ AST/STBC Scheme

Motivated by the N-out-of-$(N+1)$ AST - a particular case of the N-out-of-M AST where $M = (N+1)$ - proposed in [Katz et al., 2001], we propose here the improved N-out-of-$(N+1)$ AST for N-out-of-$(N+1)$ AST for wireless channels using STBCs (with coherent detection). For simplicity, we refer to our proposed AST as the $(N+1, N; K)$ AST/STBC. This notation means that N Tx antennas are selected from $(N+1)$ Tx antennas, while all K Rx antennas are used without selection. The $(N+1, N; K)$ AST/STBC provides the same BER performance as the N-out-of-$(N+1)$ AST in [Katz et al., 2001], but shortens the time required to process the feedback information.

It is noted that, for $M > (N+1)$, although the proposed AST/DSTBC schemes shorten the time required to process the feedback information, there exists degradation of the bit error property of the systems, compared to the case of the N-out-of-$(N+1)$ AST proposed in [Katz et al., 2001]. Therefore, we concentrate our consideration only on the case where $M = (N+1)$ in this section. To consider the principle of the proposed technique, let us consider the diagram shown in Fig. 6.1. The system comprises $(N+1)$ Tx antennas, including N default Tx antennas and one standby Tx antenna, and K Rx antennas. Without loss of generality, we number the N default Tx antennas by the indices from 1 to N. The standby Tx antenna is indexed by $(N+1)$. We denote the transmission coefficients to be h_{ij}, for $i = 1, \ldots, K$ and $j = 1, \ldots, N+1$, corresponding to the channel between the jth Tx antenna and the ith Rx antenna. Let $\xi_j \equiv \sum_{i=1}^{K} |h_{ij}|^2$ for $j = 1, \ldots, N+1$.

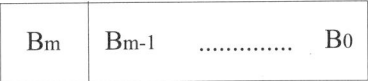

Figure 6.2. The proposed structure of the feedback information for channels using STBCs.

The antenna selection criterion in the proposed $(N + 1, N; K)$ AST/STBC is as follows. The receiver searches for the minimum value ξ_{min} among $\{\xi_1, ..., \xi_N, \xi_{N+1}\}$. For ease of exposition, we assume that $\xi_{min} \equiv \xi_k$, for $k = 1, ..., N + 1$. Then, the receiver checks whether the index k is equal to $(N+1)$ which is corresponding to the standby Tx antenna. If $k \equiv (N+1)$, then the Tx antennas which should be selected are $\{1, 2, ..., N\}$. These antennas are all the default Tx antennas. Otherwise, the kth Tx antenna is replaced by the standby antenna (the $(N + 1)^{th}$ Tx antenna) and the Tx antennas, which should be selected, are $\{1, 2, ..., k - 1, N + 1, k + 1, ..., N\}$.

Hence, the standby antenna is used when ξ_{N+1} corresponding to the standby antenna is not the worst among $(N + 1)$ values $\{\xi_1, ..., \xi_N, \xi_{N+1}\}$. It is easy to realize that the proposed technique provides the same bit error property as the N-out-of-$(N + 1)$ AST proposed in [Katz et al., 2001] since both techniques choose the N best Tx antennas from $(N + 1)$ available antennas to transmit signals.

Next we consider the structure of the feedback information and the delay required to process the feedback information at the transmitter. For simplicity, we assume that the feedback loop is error-free. We propose the structure of the feedback information used for selecting Tx antennas as presented in Fig. 6.2.

In this figure, bit B_m is used to indicate whether the transmitter has to replace the kth Tx antenna with the standby antenna. The bit B_m is zero if the answer is no and B_m is unity otherwise. The m following bits indicate which Tx antenna among N default Tx antennas should be replaced by the standby antenna. It is easy to realize that $m = \lceil \log_2 N \rceil$. With this structure, the transmitter considers the bit B_m at first. As soon as it realizes that $B_m = 0$, the rest of the feedback information is not necessarily processed [1]. The transmitter will transmit signals via the default Tx antennas $\{1, 2, ..., N\}$. If $B_m = 1$, the transmitter uses the m following bits $B_{m-1}, ..., B_0$ to recognize which default Tx antenna should be replaced by the standby antenna. Thereby, the delay for processing the feedback information is reduced. The flow chart of the proposed technique is presented in Fig. 6.3.

---

[1]Theoretically, there is no need to transmit m bits $B_{m-1}, ..., B_0$ in the case $B_m = 0$

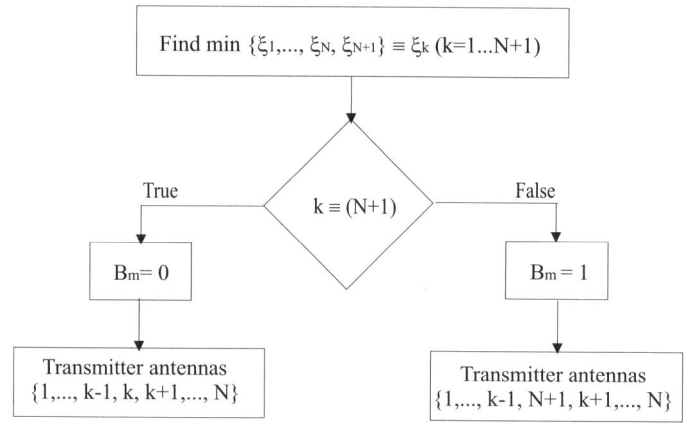

Figure 6.3. The flow chart of the proposed $(N + 1, N; K)$ AST/SBTC scheme.

In order to estimate the time benefit achieved by the proposed technique, we compare the average processing time required for our proposed $(N + 1, N; K)$ AST/STBC and for the N-out-of-$(N + 1)$ AST proposed in [Katz et al., 2001]. Although, there is a fact that the time required to process the feedback information does not necessarily linearly increase with the number of feedback bits, it is easier to calculate the time benefit of the proposed technique if the average processing time is assumed to increase linearly with the number of feedback bits. Obviously, the result we derive as follows is only aimed at providing the readers with the estimation of the average processing time saved by our technique in comparison with that of the technique proposed in [Katz et al., 2001].

Due to the assumption that the transmission coefficients h_{ij} are i.i.d complex Gaussian random variables, the events in which ξ_k, for $k = 1, \ldots, (N + 1)$, is the smallest value among $(N + 1)$ values $\{\xi_1, \ldots, \xi_{N+1}\}$ are equiprobable. In other words, the probability of the event in which ξ_{N+1} is the smallest value among $(N + 1)$ values $\{\xi_1, \ldots, \xi_{N+1}\}$ is $\frac{1}{(N+1)}$. In this case, $B_m = 0$ and the transmitter has to process only one bit (bit B_m). The probability of the event in which ξ_{N+1} is *not* the smallest value among $(N + 1)$ values $\{\xi_1, \ldots, \xi_{N+1}\}$ is $1 - \frac{1}{(N+1)} = \frac{N}{N+1}$. The transmitter now has to process $(m+1) = (1 + \lceil \log_2 N \rceil)$ bits. Let t be the average processing time for one feedback bit, then the average time required to process feedback information in our method is

$$\tau_1 = \frac{1}{N + 1} t + \frac{N}{N + 1} (1 + \lceil \log_2 N \rceil) t$$

On the other hand, in the N-out-of-$(N + 1)$ AST proposed in [Katz et al.,

Table 6.1. The average processing time reduction of the proposed $(N+1, N; K)$ AST/STBC technique

N	$\frac{\triangle \tau}{\tau_2}$
2	16.67 %
4	13.33 %
8	8.33 %

2001], the transmitter always has to process

$$\left\lceil \log_2 \binom{N+1}{N} \right\rceil = \left\lceil \log_2(N+1) \right\rceil$$

bits.

Therefore, the average processing time is

$$\tau_2 = \left\lceil \log_2(N+1) \right\rceil t$$

It is easy to realize that when N is the power of 2, for instance $N = 2, 4, 8$, one has

$$1 + \left\lceil \log_2 N \right\rceil = \left\lceil \log_2(N+1) \right\rceil \tag{6.3}$$

Hence, the relative reduction of the average processing time between two techniques is

$$\frac{\triangle \tau}{\tau_2} = \frac{\tau_2 - \tau_1}{\tau_2} = \frac{\left\lceil \log_2 N \right\rceil}{(N+1)\left\lceil \log_2(N+1) \right\rceil} \tag{6.4}$$

The formula (6.3) shows that, when $B_m = 1$, our $(N+1, N; K)$ AST/STBC uses the same number of feedback bits for selecting Tx antennas as the N-out-of-$(N+1)$ AST proposed in [Katz et al., 2001], while the formula (6.4) shows that the time required to process the feedback information in the former is shorter than that in the later. The average processing time reduction for some particular values is given in Table 6.1. From this table, we realize that the proposed technique allows the transmitter to reduce the time required to process the feedback information, especially for $N = 2$, such as when the Alamouti STBC [Alamouti, 1998] is utilized.

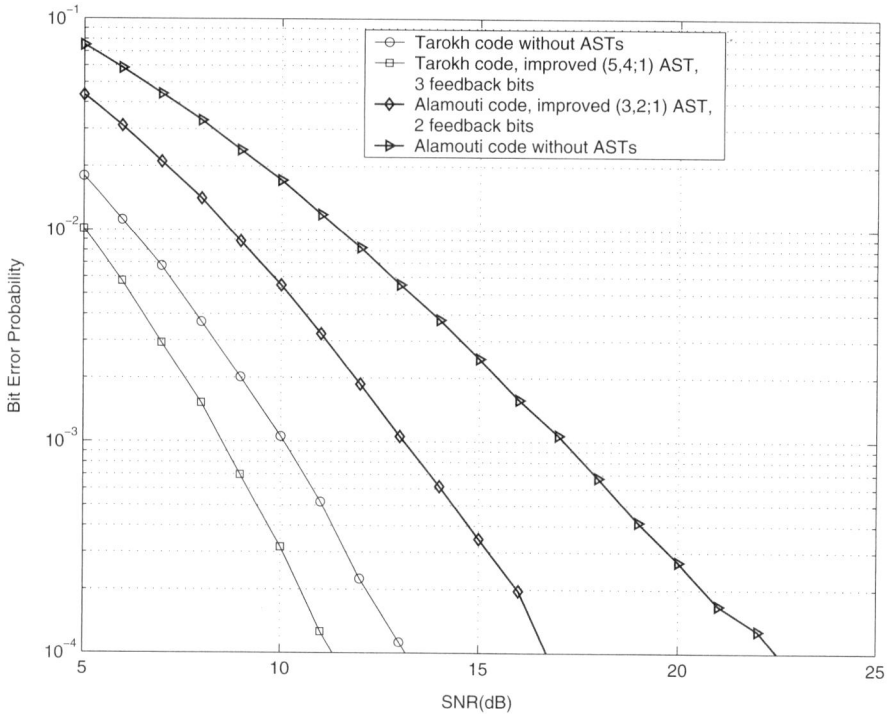

Figure 6.4.   BER vs. SNR for the Alamouti code and the Tarokh code $\mathbf{G}4$ [Tarokh et al., 1999a] with and without antenna selection.

Finally, we consider the capacity limitation of uplink channels used for transmit diversity purposes in WCDMA (Wideband Code Division Multiple Access) mobile communication systems. The feedback information used for transmit diversity purposes is usually transmitted in the D bit field of the Feedback Information field in the uplink Dedicated Physical Control Channel (DPCCH). According to the standard of WCDMA systems [3GPP, 2000], the rate of the feedback information used for closed loop mode transmit diversity is limited to 1500 bits/sec (the maximal number of feedback bits per time slot is 1). Hence, $m + 1$ bits used for selecting Tx antennas are transmitted during $m + 1$ consecutive slots. In order to keep the transmitter updated with a small delay on the best channels, we suggest that the number of feedback bits should not be greater than 4 (corresponding to $N=8$ Tx antennas).

6.2.3   Simulation Results

In this section, we compare the bit error probability of STBCs transmitted via channels with and without our proposed AST. A system containing 1 Rx

Table 6.2. Comparison between the proposed $(N + 1, N; K)$ AST/STBC scheme and the technique proposed in [Katz et al., 2001].

Criteria	Proposed technique vs. N-out-of-(N + 1) AST in [Katz et al., 2001]
Bit error performance	the same
Number of feedback bits	at most, the same
Number of transmitter antennas	the same
Average processing time	shorter (16.7% for $N = 2$)

antenna is considered. The signal-to-noise ratio SNR is defined by the ratio between the total received power of the transmitted signals and the power of noises at the receiver during each STS.

Fig. 6.4 shows the bit error probability of the Alamouti code ($N = 2$) modulated by a QPSK constellation with our (3,2;1) AST/STBC scheme. It can be seen from the figure that the SNR gain achieved by our technique is 5.5 dB at BER=10^{-4} compared to the case without antenna selection.

Fig. 6.4 also presents the bit error probability of the rate-1/2 STBC proposed by V. Tarokh et al. [Tarokh et al., 1999a] ($N = 4$) with and without our proposed (5,4;1) AST/STBC scheme. The SNR advantage gained by our (5,4;1) AST/STBC scheme in this case is 2 dB at BER=10^{-4}.

A comparison between our technique and the technique proposed in [Katz et al., 2001] is presented in Table 6.2.

6.3 Transmitter Diversity Antenna Selection Techniques for Wireless Channels Utilizing DSTBCs

6.3.1 Reviews on DSTBCs

In this section, we review the conventional Differential Space-Time Block Codes (DSTBCs) mentioned in the literature and provide some remarks on the time-varying Rayleigh fading channels where DSTBCs can be practically used. This section is indispensable in order for the readers to comprehend what has been modified in the transmission procedures of DSTBCs in our proposed ASTs. It is also vital for the readers to recognize the underlying requirement of all conventional DSTBCs that the channel coefficients must be constant during at least two consecutive code blocks. We also show here in which scenario DSTBCs (with differential detection) should be used, rather than STBCs (with coherent detection).

6.3.1.1   Conventional DSTBCs without Diversity Antenna Selection

DSTBCs are useful for channels where fading changes so fast that the transmission of a *large* overhead of training signals is either impractical or uneconomical. DSTBCs have been considered intensively and a number of DSTBCs have been proposed in the literature such as [Chen et al., 2003d], [Ganesan and Stoica, 2002a], [Ganesan and Stoica, 2002b], [Hochwald and Sweldens, 2000], [Hughes, 2000], [Tarokh and Jafarkhani, 2000], [Vucetic and Yuan, 2003].

Let us consider the *square, unitary* DSTBC proposed by Ganesan et al. in [Ganesan and Stoica, 2002b] as an example. We consider a system with N Tx antennas and K Rx antennas. Let $\mathbf{R}_t$, $\mathbf{A}$, $\mathbf{N}_t$ be the ($K \times N$)-sized matrices of received signals at time t, channel coefficients between the Rx and Tx antennas, and noise at the Rx antennas, respectively. It is noted that all these matrices have the same sizes $K \times N$ since we are considering the *square*, order-N DSTBC. The $\kappa\eta^{th}$ element of $\mathbf{A}$, namely $a_{\kappa\eta}$, is the channel coefficient of the path between the ηth Tx antenna and the κth Rx antenna. Channel coefficients are assumed to be i.i.d. complex, zero-mean Gaussian random variables. Noises are assumed to be i.i.d. complex Gaussian random variables with the distribution $\mathcal{CN}(0, \sigma^2)$.

Let $\{s_j\}_{j=1}^p = \{s_j^R + is_j^I\}_{j=1}^p$ (where $i^2=-1$, s_j^R and s_j^I are the real and imaginary parts of s_j, respectively) be the set of p symbols, which are derived from a *unit* power signal constellation S and transmitted in the tth block. Consequently, each symbol has a unit energy, i.e. $|s_j|^2 = 1$.

We define a square matrix $\mathbf{Z}_t = \frac{1}{\sqrt{p}} \sum_{j=1}^p (\mathbf{X}_j s_j^R + i\mathbf{Y}_j s_j^I)$, where the square, order-N weighting matrices $\{\mathbf{X}_j\}_{j=1}^p$ and $\{\mathbf{Y}_j\}_{j=1}^p$ are orthogonal themselves and satisfy the permutation property. These weighting matrices are considered as the amicable orthogonal designs (AODs). The background knowledge of AODs can be found in [Geramita and Seberry, 1979]. The coefficient $\frac{1}{\sqrt{p}}$ is to guarantee that $\mathbf{Z}_t$ is a unitary matrix, i.e., $\mathbf{Z}_t\mathbf{Z}_t^H = \mathbf{I}$.

For illustration, the Alamouti DSTBC corresponding to $N = 2$ is defined as

$$\mathbf{Z}_t = \frac{1}{\sqrt{2}} \begin{bmatrix} s_1 & s_2 \\ -s_2^* & s_1^* \end{bmatrix} \tag{6.5}$$

A DSTBC corresponding to $N = 4$ is given below:

$$\mathbf{Z}_t = \frac{1}{\sqrt{3}} \begin{bmatrix} s_1 & s_2 & s_3 & 0 \\ -s_2^* & s_1^* & 0 & s_3 \\ -s_3^* & 0 & s_1^* & -s_2 \\ 0 & -s_3^* & s_2^* & s_1 \end{bmatrix} \tag{6.6}$$

The transmission starts with an initial, identity, order-N matrix $\mathbf{W}_0 = \mathbf{I}_N$ carrying *no* information. This transmission is referred to as the *initial transmission*. The matrix transmitted at time t $(t = 1, 2, 3 \ldots)$ is given by

$$\mathbf{W}_t = \mathbf{W}_{t-1}\mathbf{Z}_t \tag{6.7}$$

As $\mathbf{Z}_t$ is a unitary matrix, the matrix $\mathbf{W}_t$ is also a unitary one. The model of the channel at time t, for $t = 0, 1, 2 \ldots, (t = 0$ means the initial transmission) is

$$\mathbf{R}_t = \mathbf{A}\mathbf{W}_t + \mathbf{N}_t \tag{6.8}$$

In all proposed conventional DSTBCs [Chen et al., 2003d], [Ganesan and Stoica, 2002a], [Ganesan and Stoica, 2002b], [Hochwald and Sweldens, 2000], [Hughes, 2000], [Tarokh and Jafarkhani, 2000], [Vucetic and Yuan, 2003], the channel coefficients must be constant during *at least* two adjacent code blocks, i.e. constant during *at least* $2N$ symbol time slots (STSs). It means that if the channel coefficient matrix $\mathbf{A}$ is assumed to be constant over two consecutive blocks $t - 1$ and t, the maximum likelihood (ML) detector for the symbols $\{s_j\}_{j=1}^p$ is calculated as follows [Ganesan, 2002], [Ganesan and Stoica, 2002b]:

$$\{\hat{s}_j\}_{j=1}^p = Arg\left\{ \max_{\{s_j\}, s_j \in S} Re\{tr(\mathbf{R}_t^H \mathbf{R}_{t-1}\mathbf{Z}_t)\} \right\} \tag{6.9}$$

where $Arg\{.\}$ denotes the argument operation, $tr(.)$ denotes the trace operation, $Re\{.\}$ and $Im\{.\}$ denote the real and the imaginary parts of the argument, respectively.

If we denote T_c to be the average coherence time of the channel which represents the time-varying nature of the channel, then the channel is considered to be constant during this time. After each duration T_c, the transmitter restarts the transmission and transmits a new initial block $\mathbf{W}_0$ followed by other code blocks $\mathbf{W}_t$ $(t = 1, 2, 3 \ldots)$. This procedure is repeated until all data are transmitted.

Due to the orthogonality of DSTBCs, the transmitted symbols are decoded separately, rather than jointly. Therefore, if we denote

$$D_j = Re\{tr(\mathbf{R}_t^H \mathbf{R}_{t-1}\mathbf{X}_j)\} + iRe\{tr(\mathbf{R}_t^H \mathbf{R}_{t-1}i\mathbf{Y}_j)\} \tag{6.10}$$

then the ML detector for the symbol s_j is (see Appendix B of this book or [Tran and Wysocki, 2003], [Tran and Wysocki, 2004])

$$\hat{s}_j = Arg\left\{ \max_{s_j \in S} Re\{D_j^* s_j\} \right\} \tag{6.11}$$

where D_j^* is the conjugate of D_j.

Expressions (6.10) and (6.11) show that the detection of the symbol s_j is carried out without the knowledge of channel coefficients. Particularly, the symbol s_j can be decoded by using the received signal blocks, $\mathbf{R}_{t-1}$ and $\mathbf{R}_t$, in two consecutive transmissions $(t-1)^{th}$ and t^{th}, *provided that the channel coefficients are constant during these two consecutive code blocks* (otherwise, the decoding expressions (6.9) and (6.11) will not hold).

The requirement that the channel coefficients must be constant during at least two consecutive code blocks can be relaxed if linear prediction is used at the receiver. In this scenario, the receiver uses multiple previously received code blocks $\mathbf{R}_{t-1}, \mathbf{R}_{t-2}$, etc. to predict the relation between the current channel coefficient matrix, say $\mathbf{A}_t$, and the previous channel coefficient matrices. This approach has been mentioned in [Song and Xia, 2003]. Certainly, a penalty of this approach is the complexity of the receiver structure.

It has been proved in our papers [Tran and Wysocki, 2003], [Tran and Wysocki, 2004] that all conventional DSTBCs (without ASTs) provide full diversity of order NK, where N and K are the number of Tx and Rx antennas, respectively, provided that the DSTBCs have a full rank.

6.3.1.2   Remarks on the Time-Varying Rayleigh Fading Channels

According to the frequency of channel coefficient changes, Rayleigh flat fading channels can be classified into three typical scenarios, i.e. *fast, block* and *quasi-static* (or *slow*) Rayleigh fading channels, which are usually examined in practice and present the most common, realistic propagation conditions. Readers should refer to Section 2.2.3 of this book or [Telatar, 1999], [Vucetic and Yuan, 2003] (pp. 13) for more details.

Because the channel coefficients must be *constant during, at least, two consecutive code blocks*, in all conventional DSTBCs mentioned in the literature, the channels are considered as *block* fading channels, although the coherence time T_c of the channels in the case of DSTBCs (with differential detection) can be *much shorter* than that in the case of STBCs (with coherent detection).

To illustrate, for the case of the Alamouti DSTBC, the channel coefficients must be constant during at least 4 STSs. During the first two STSs, the initial, order-2, identity matrix $\mathbf{I}_2$ which carries no information is transmitted. During the next two STSs, the Alamouti code carrying 2 symbols is transmitted. This note clarifies how fast fading channels may change when DSTBCs are utilized. Certainly, longer coherence time results in a more efficient utilization of the channel using DSTBCs.

We give 2 examples of block Rayleigh fading channels where STBCs (coherent detection) or DSTBCs (differential detection) can be used.

EXAMPLE 6.3.1 *We consider here the scenario where the Alamouti STBC with coherent detection can be used for the cellular mobile system with the carrier frequency $F_c = 900$ MHz. Speed of the mobile user is $v = 5$ km/h (walking speed) and each STS is assumed to be $T_s = 0.125$ ms (equivalently, the baud rate is $F_s = 8$ Kbaud). Denote $c = 3.10^8$ m/s to be the speed of light. Hence, the maximum Doppler frequency is calculated as*

$$f_m = vF_c/c = 4.17 \ Hz$$

The average coherence time T_c of the channel is estimated by the following empirical expression [Rappaport, 2002] (pp. 204):

$$T_c = \frac{0.423}{f_m} = 101.52 \ ms$$

It means that the channel coefficients can be considered to be constant during almost $T_c/T_s \approx 812$ consecutive STSs, i.e., approximately 406 consecutive Alamouti code blocks. In this case, the channel coefficients change so slowly that a large overhead of multiple training signals can be transmitted. In other words, STBCs with coherent detection are preferred than DSTBCs with differential detection.

EXAMPLE 6.3.2 *We consider here another scenario where the Alamouti DSTBC with differential detection can be used for the cellular mobile system with the carrier frequency $F_c = 900$ MHz. Speed of the mobile user is $v = 60$ km/h (vehicular speed) and each STS is assumed to be $T_s = 0.5$ ms corresponding to the baud rate $F_s = 2$ Kbaud. The maximum Doppler frequency is then calculated as*

$$f_m = vF_c/c = 50 \ Hz$$

Similarly, the average coherence time T_c of the channel is estimated as [Rappaport, 2002] (pp. 204)

$$T_c = \frac{0.423}{f_m} = 8.46 \ ms$$

It means that the channel coefficients can be considered to be constant during $T_c/T_s \approx 16$ consecutive STSs, i.e., 8 consecutive Alamouti code blocks. The channel is a block Rayleigh fading one where DSTBCs can be employed. In this case, it is either impractical or uneconomical to use STBCs with coherent detection since the coherence time is too short to transmit multiple training symbols in order for the receiver to estimate exactly the channel coefficients.

6.3.2   Definitions, Notations and Assumptions

To facilitate future considerations, we introduce here the following notations:

DEFINITION 6.3.1 *F is defined as an order-N operation on M non-negative, real numbers $\{\varepsilon_1, \ldots, \varepsilon_M\}$ where the N indices ($N < M$) corresponding to the N largest values out of M values $\{\varepsilon_1, \ldots, \varepsilon_M\}$ are selected. We denote this operation as $F_N(\varepsilon_1, \ldots, \varepsilon_M)$. The output of the operation F is the set of N indices which is denoted by $\hat{\mathcal{I}}_N$.*

EXAMPLE 6.3.3 *Consider $M = 3$, $N = 2$, $\varepsilon_1 = 10$, $\varepsilon_2 = 20$ and $\varepsilon_3 = 30$. We have*

$$\hat{\mathcal{I}}_2 = F_2(\varepsilon_1, \varepsilon_2, \varepsilon_3) = \{2, 3\}$$

The elements of the set $\hat{\mathcal{I}}_2$ are the indices of ε_2 and ε_3, which are, in turn, the two largest values among $\{\varepsilon_1, \varepsilon_2, \varepsilon_3\}$.

DEFINITION 6.3.2 *We define the $(M, N; K, L)$ AST/DSTBC scheme to be the transmitter and receiver diversity antenna selection technique for channels using DSTBCs with differential detection where N Tx antennas are selected out of M Tx antennas ($N < M$), while L Rx antennas are selected out of K Rx antennas ($L < K$) for transmission.*

Given that notation, the $(M, N; K)$ AST/DSTBC scheme refers to the *transmitter* diversity antenna selection technique for channels using DSTBCs with *differential detection*, where N Tx antennas are selected out of M Tx antennas for transmission, while all K Rx antennas are used without selection. Similarly, the $(M; K, L)$ AST/DSTBC scheme refers to the *receiver* diversity AST for channels using DSTBCs, where L Rx antennas are selected out of K Rx antennas for reception, while M Tx antennas are used without selection.

In the chapter, we mainly focus on the transmitter diversity AST, i.e., the $(M, N; K)$ AST/DSTBC schemes. We sometimes compare the proposed $(M, N; K)$ AST/DSTBC schemes with the respective schemes in channels using STBCs with *coherent detection*. Hence, similarly, we use the notation $(M, N; K)$ AST/STBC to refer to the transmitter diversity AST for channels using STBCs with *coherent detection*.

We consider here some assumptions as given below

ASSUMPTION 6.3.3 *The channel coefficients between the Tx and Rx antennas are assumed to be i.i.d. complex, zero-mean Gaussian random variables. Noises are assumed to be i.i.d. complex Gaussian random variables with the distribution $\mathcal{CN}(0, \sigma^2)$. These assumptions are applicable when the Tx and Rx* *antennas are sufficiently separated from each other (by a multiple of half of the wavelength). The scenario where the antennas are correlated will be examined in the next chapter.*

ASSUMPTION 6.3.4 *Although channels with differential detection change faster than those with coherent detection, so that the transmission of a large overhead consisting of multiple training signals is uneconomical (and, consequently, the utilization of DSTBCs is useful), we make a reasonable assumption that it is possible to transmit a few feedback bits (for each channel coherence time T_c) from the receiver to the transmitter via a feedback channel with a certain feedback error rate. The feedback error rate is typically assumed to be 4% to 10% [Choi et al., 2002], [Hamalainen and Wichman, 2002].*

Finally, we would like to stress that employment of more than two Rx antennas at the receiver is uneconomical due to the tiny size of the receivers, such as the hand-held mobile phones in cellular mobile systems. Hence, receiver diversity antenna selection is not considered in this chapter, although the generalization of the proposed techniques for the receiver diversity antenna selection is straightforward.

6.3.3   Basis of Transmitter Antenna Selection for Channels Using DSTBCs

As stated earlier, antenna selection for channels using STBCs has been intensively mentioned in the literature, such as in [Gore and Paulraj, 2002]. Antenna subset selection in channels using STBCs with coherent detection can be done with either exact channel knowledge (ECK) or statistical channel knowledge (SCK) [Gore and Paulraj, 2002]. In ECK-based selection, the *instantaneous* error probability of the system is minimized. In SCK-based selection, however, the *average* probability of errors is minimized over all possible channel realizations. Antenna selection algorithms have also been derived in [Gore and Paulraj, 2002] for both scenarios.

However, these algorithms are derived only for channels using STBCs where channel coefficients are known at the receiver. Antenna selection for channels using DSTBCs was not mentioned in [Gore and Paulraj, 2002]. In general, ASTs for channels using DSTBCs have not been well examined yet. Therefore, in the rest of this chapter, we propose some antenna selection algorithms for such channels.

Let us consider the unitary DSTBCs mentioned in Section 6.3.1.1 for instance. It is shown in [Ganesan, 2002] (Eq. (5.30)), [Ganesan and Stoica,

2002a], [Tran and Wysocki, 2003] (Eq. (11)), or [Tran and Wysocki, 2004] (Eq. (9)) that the SNR of the statistic D_j in (6.10) is approximately

$$SNR_{diff} \approx \frac{\|\mathbf{A}\|_F^2}{2p\sigma^2} = \frac{tr(\mathbf{A}^H \mathbf{A})}{2p\sigma^2} = \frac{\sum_{\eta=1}^{N} \left[\sum_{\kappa=1}^{K} |a_{\kappa\eta}|^2 \right]}{2p\sigma^2} \quad (6.12)$$

where $\|\mathbf{A}\|_F$ is the Frobenius norm of the channel coefficient matrix $\mathbf{A}$. Clearly, the SNR in (6.12) has $2NK$ freedom degrees. As a result, the considered unitary DSTBC provides full spatial diversity of order NK, provided that the DSTBC has a full rank [Tran and Wysocki, 2003], [Tran and Wysocki, 2004].

Let $\xi_\eta \triangleq \sum_{\kappa=1}^{K} |a_{\kappa\eta}|^2$, for $\eta = 1, \ldots, N$. We can rewrite SNR_{diff} as follows:

$$SNR_{diff} \approx \frac{\sum_{\eta=1}^{N} \xi_\eta}{2p\sigma^2} \quad (6.13)$$

It is obvious that greater values of ξ_ηs result in the greater SNR_{diff}.

We now consider a system containing M Tx antennas ($M > N$) and K Rx antennas. We would like to select the N best Tx antennas out of M Tx antennas so that SNR_{diff} is maximized. From (6.12) or (6.13), to maximize SNR_{diff}, we need to maximize $\|\mathbf{A}\|_F^2$. Equivalently, the N *first maximum values* out of M values $\{\xi_1, \xi_2, \ldots, \xi_M\} = \left\{ \sum_{\kappa=1}^{K} |a_{\kappa 1}|^2, \sum_{\kappa=1}^{K} |a_{\kappa 2}|^2, \ldots, \sum_{\kappa=1}^{K} |a_{\kappa M}|^2 \right\}$ must be selected. In other words, the indices of the N best Tx antennas are selected by the following antenna selection criterion:

$$\hat{\mathcal{I}}_N = \Gamma_N \left(\xi_1, \ldots, \xi_M \right)$$
$$= \Gamma_N \left(\sum_{\kappa=1}^{K} |a_{\kappa 1}|^2, \sum_{\kappa=1}^{K} |a_{\kappa 2}|^2, \ldots, \sum_{\kappa=1}^{K} |a_{\kappa M}|^2 \right) \quad (6.14)$$

Again, note that *transmitter* diversity antenna selection, rather than *receiver* diversity antenna selection, is examined in this chapter. All K Rx antennas are used without antenna selection.

The selection criterion in (6.14) is applicable only when the channel coefficients are perfectly known at the receiver. This scenario is realistic when the channel changes so slowly that a large overhead containing *multiple* training signals can be transmitted. This scenario is commonly examined in channels using STBCs with *coherent detection*. The ASTs are referred to as the $(M, N; K)$ AST/STBC schemes which have been intensively considered in the literature

[Chen, 2004], [Chen et al., 2003b], [Chen et al., 2003c], [Gore and Paulraj, 2001], [Gore and Paulraj, 2002], [Katz et al., 2001], [Xiaofeng et al., 2001].

As opposed to *coherent detection*, in channels using DSTBCs with *differential detection*, channel coefficients change faster so that the transmission of *numerous* training signals is either impractical or uneconomical, and consequently, the channel coefficients cannot be exactly known at the receiver. Therefore, the antenna selection criterion in (6.14) cannot be directly applied to channels using DSTBCs with *differential detection*. However, we will show that this criterion can be *modified* to apply to such channels.

Particularly, we will prove later in this chapter that, at high SNRs, the statistical properties, i.e. means and variances, of the received signals $r_{0\kappa\eta}$s, which are the elements of the initial received matrix $\mathbf{R}_0$ during the initial transmission, are similar to those properties of the channel coefficients $a_{\kappa\eta}$s. As a result, at high SNRs, maximizing $\|\mathbf{R}_0\|_F^2$ tends to be the same as maximizing $\|\mathbf{A}\|_F^2$.

Based on this observation, we propose the *modified* antenna selection scheme for channels using DSTBCs. The transmitter selects Tx antennas leaning on the comparison, which is carried out (once during each channel coherence time T_c) at the receiver, between the powers received by all K Rx antennas from different Tx antennas during the initial transmission (when the first block $\mathbf{W}_0$ is transmitted).

If we denote $\hat{\mathcal{I}}_N$ to be the set of the N indices of the Tx antennas which should be selected, then the *modified* antenna selection criterion for channels using DSTBCs is

$$
\begin{aligned}
\hat{\mathcal{I}}_N &= F_N\left(\chi_1, \ldots, \chi_M\right) \\
&= F_N\left(\sum_{\kappa=1}^{K} |r_{0\kappa1}|^2, \sum_{\kappa=1}^{K} |r_{0\kappa2}|^2, \ldots, \sum_{\kappa=1}^{K} |r_{0\kappa M}|^2\right)
\end{aligned}
$$

where $\chi_\eta \triangleq \sum_{\kappa=1}^{K} |r_{0\kappa\eta}|^2$ for $\eta = 1, \ldots, M$. From the mathematical point of view, the N selected antennas are corresponding to the N columns out of M columns of the *initial received matrix* with the maximum Frobenius norms. Clearly, channel coefficients are not required to be known at the receiver in this case. This modified selection criterion is mentioned in more details in the so-called *general* $(M, N; K)$ AST/DSTBC scheme proposed in the next section.

6.3.4　The General $(M, N; K)$ AST/DSTBC Scheme for Channels Utilizing DSTBCs

In this section, we generalize our AST/DSTBC schemes proposed in [Tran and Wysocki, 2003], [Tran and Wysocki, 2004] for channels using DSTBCs with *arbitrary* numbers of Tx and Rx antennas.

Let us consider a system containing M Tx antennas and K Rx antennas using the unitary, square, order-N DSTBCs ($N < M$) proposed by Ganesan et al. [Ganesan and Stoica, 2001], [Ganesan and Stoica, 2002b]. Note that the proposed ASTs mentioned here are also applicable to any conventional DSTBC, regardless of being unitary or not.

In the following analysis, the bold, lower case letters denote vectors and the bold upper case letters denote matrices, while the normal letters denote scalars. For simplicity, we omit the superscripts indicating the different coherence durations T_cs of the channel when a certain coherence duration is being considered. The superscripts are only used when we consider different coherence durations simultaneously.

The *general* $(M, N; K)$ AST/DSTBC is proposed as follows:

- At the beginning of transmission during each coherence time T_c, the transmitter sends an initial block $\tilde{\mathbf{W}}_0 = \mathbf{I}_M$ via M Tx antennas, rather than sending an initial block $\mathbf{W}_0 = \mathbf{I}_N$ via N Tx antennas like in all conventional DSTBCs. This transmission is referred to as the *initial transmission*.

We note the change in the size of matrices, compared to (6.8), by using the tilde mark for matrices as below

$$\tilde{\mathbf{W}}_0 = \mathbf{I}_M$$

$$\tilde{\mathbf{A}} = \begin{bmatrix} \mathbf{a_1} & \mathbf{a_2} & \cdots & \mathbf{a_M} \end{bmatrix}$$

$$\tilde{\mathbf{N}}_0 = \begin{bmatrix} \mathbf{n_{01}} & \mathbf{n_{02}} & \cdots & \mathbf{n_{0M}} \end{bmatrix}$$

where $\mathbf{a_j}$ ($j=1\ldots M$) is the column vector of the channel coefficients a_{ij} ($i = 1, \ldots, K$) corresponding to the channel from the jth Tx antenna to the ith Rx antenna, i.e, $\mathbf{a_j} = (a_{1j}, \ldots, a_{Kj})^T$, and $\mathbf{n_{0j}}$ is the noise affecting these channels during the initial transmission, i.e., $\mathbf{n_{0j}} = (n_{01j}, \ldots, n_{0Kj})^T$. Here, the superscript T denotes the transposition operation.

- The receiver determines the initial received matrix $\tilde{\mathbf{R}}_0$ during the initial transmission as given below

$$
\begin{aligned}
\tilde{\mathbf{R}}_0 &= \tilde{\mathbf{A}}\tilde{\mathbf{W}}_0 + \tilde{\mathbf{N}}_0 \\
&= \tilde{\mathbf{A}}\mathbf{I}_M + \tilde{\mathbf{N}}_0 \\
&= \begin{bmatrix} \mathbf{a}_1 + \mathbf{n}_{01} & \mathbf{a}_2 + \mathbf{n}_{02} & \cdots & \mathbf{a}_M + \mathbf{n}_{0M} \end{bmatrix} \\
&= \begin{bmatrix} \mathbf{r}_{01} & \mathbf{r}_{02} & \cdots & \mathbf{r}_{0M} \end{bmatrix}
\end{aligned}
\tag{6.15}
$$

where

$$
\begin{aligned}
\mathbf{r}_{0j} &\overset{\Delta}{=} \mathbf{a}_j + \mathbf{n}_{0j} \\
&= \left(a_{1j} + n_{01j}, \ldots, a_{Kj} + n_{0Kj} \right)^T, \quad j = 1 \ldots M
\end{aligned}
\tag{6.16}
$$

- From the initial, received matrix $\tilde{\mathbf{R}}_0$, the receiver determines semi-blindly the N best channels by comparing M terms:

$$
\chi_j = \|\mathbf{r_{0j}}\|_F^2 = \sum_{i=1}^{K} |r_{0ij}|^2 = \sum_{i=1}^{K} |a_{ij} + n_{0ij}|^2
\tag{6.17}
$$

for $j = 1 \ldots M$, to search for the first N maximum values.

Note that χ_j is the total power received by all K Rx antennas from the jth Tx antenna in the initial transmission. Therefore, intuitively, the receiver searches for the first N largest values of received powers, among M values.

The antenna selection criterion can be mathematically expressed as

$$
\begin{aligned}
\hat{\mathcal{I}}_N &= F_N(\chi_1, \ldots, \chi_M) \\
&= F_N\left(\|\mathbf{r}_{01}\|_F^2, \ldots, \|\mathbf{r}_{0M}\|_F^2\right)
\end{aligned}
\tag{6.18}
$$

where $\hat{\mathcal{I}}_N$ denotes the set of N indices of the Tx antennas which should be selected. From the mathematical point of view, the receiver searches for the N columns having maximum Frobenius norms among M columns of the matrix $\tilde{\mathbf{R}}_0$.

Without loss of generality, we assume here that these maximum values are corresponding to the *first* N elements of the matrix $\tilde{\mathbf{R}}_0$, i.e. $\mathbf{r}_{01}, \ldots, \mathbf{r}_{0N}$. It follows that

$$
\hat{\mathcal{I}}_N = \{1, 2, \ldots, N\}
$$

▪ The receiver then carries out the two following tasks:

1 It informs the transmitter via a feedback channel to select the first N Tx antennas to transmit data.

2 It generates the matrix $\mathbf{R}_0$, which is used to decode the next code blocks, by taking the first N columns of the matrix $\tilde{\mathbf{R}}_0$, corresponding to the first N maximum values, i.e.

$$\mathbf{R}_0 = \begin{bmatrix} \mathbf{a}_1 + \mathbf{n}_{01} & \mathbf{a}_2 + \mathbf{n}_{02} & \cdots & \mathbf{a_N} + \mathbf{n_{0N}} \end{bmatrix}$$

▪ The transmitter selects the N Tx antennas indicated by the feedback information. In this case, the first N Tx antennas are selected to transmit data. The transmission (after the initial transmission) is now exactly the same as that in the system using the first N Tx antennas only.

After each coherence duration T_c, the transmitter restarts the transmission and transmits a new initial block $\tilde{\mathbf{W}}_0$ followed by other code blocks $\mathbf{W}_t$ ($t = 1, 2, 3 \ldots$). The above procedures are repeated for different coherence durations until all data are transmitted.

The transmission procedure of DSTBCs with the proposed AST is shown in Fig. 6.5. The superscripts are used to indicate the different coherence durations T_c of the channel. The code blocks $\tilde{\mathbf{W}}_0$ are transmitted via M Tx antennas in M STSs and the blocks following $\tilde{\mathbf{W}}_0$ via the N selected Tx antennas in N STSs.

From the aforementioned algorithm, we have following remarks:

REMARK 6.3.5 *At the transmitter, after the initial matrix* $\tilde{\mathbf{W}}_0$ *is transmitted, the next matrices* $\mathbf{W}_t$ *(t=1, 2, 3, . . .) can be calculated, following Eq. (6.7), by using a tacit default matrix* $\mathbf{W}_0 = \mathbf{I}_N$. *We use the term "tacit default matrix" to refer to the fact that the matrix* $\mathbf{W}_0 = \mathbf{I}_N$ *is only used at the transmitter to generate the next code blocks* $\mathbf{W}_t$ *by Eq. (6.7), rather than being actually transmitted. Due to this fact, it is also important to note that the generation of the matrices* $\mathbf{W}_t$ *does not necessarily take place after the transmitter obtains the feedback information. Instead, the next code blocks* $\mathbf{W}_t$ *are automatically generated, following Eq. (6.7), by multiplying the previous block* $\mathbf{W}_{t-1}$ *with the tacit default matrix* $\mathbf{W}_0 = \mathbf{I}_N$.

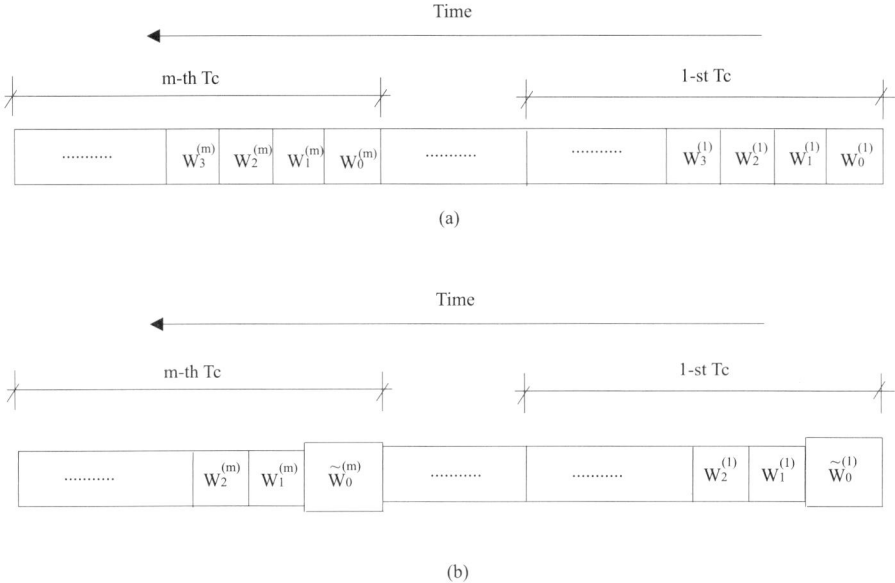

Figure 6.5. Transmission of DSTBCs (a) without and (b) with the antenna selection technique.

REMARK 6.3.6 *The number of feedback bits required to inform the transmitter about the best channels in the general* $(M, N; K)$ *AST/DSTBC is*

$$\mathcal{N} = \left\lceil \log_2 \binom{M}{N} \right\rceil \tag{6.19}$$

where $\lceil . \rceil$ *is the ceiling function.*

REMARK 6.3.7 *In all conventional DSTBCs, the initial matrices* $\mathbf{W}_0 = \mathbf{I}_N$ *are only used to initialize the transmission. Unlike the conventional DSTBCs without ASTs, in the proposed technique, the initial identity matrices* $\tilde{\mathbf{W}}_0 = \mathbf{I}_M$ *are transmitted. These matrices initialize the transmission and, simultaneously, play a role of training signals, which assist the receiver to determine semi-blindly the best channels. This is the main difference between the role of initial code blocks in DSTBCs with and without our AST.*

The sizes of $\mathbf{W}_0$ *and* $\tilde{\mathbf{W}}_0$ *are usually not much different since, in practical systems, the number of Tx antennas rarely exceeds four to five, i.e.* $N < M \leq 5$. *The difference in the sizes of* $\tilde{\mathbf{W}}_0$ *and* $\mathbf{W}_0$ *is the number of training symbols required for each coherence duration* T_c

$$\mathcal{N}_{training} = (M - N)$$

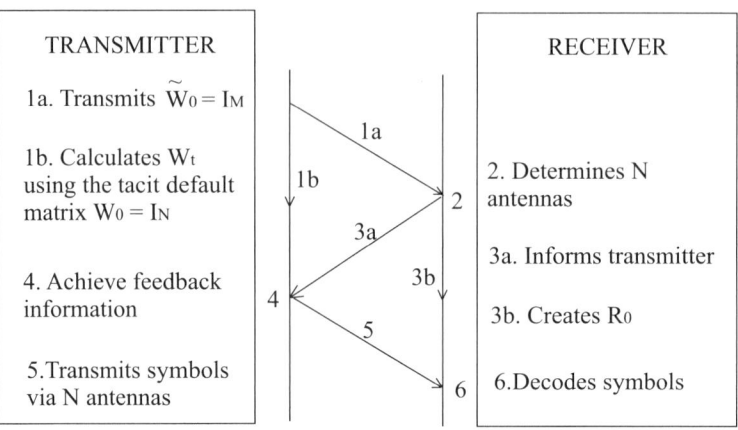

Figure 6.6. The *general* $(M, N; K)$ AST/DSTBC scheme for systems using DSTBCs.

The typical value of $\mathcal{N}_{training}$ is 1 or 2. It is important to note that the initial blocks $\mathbf{W}_0$ and $\widetilde{\mathbf{W}}_0$ are identity matrices. Hence, the penalty to achieve the statistical channel knowledge in our proposed AST is a small expansion in the size of the initial code blocks $\mathbf{W}_0$ with an extra limited transmission cost and with a slightly modified receiver structure, as opposed to the transmission of a overhead consisting of numerous training signals as in channels using STBCs (with coherent detection).

REMARK 6.3.8 *Similarly to the conventional DSTBCs without ASTs mentioned in Section 6.3.1, in our proposed technique, channel coefficients are required to be constant during at least two consecutive code blocks. Therefore, if the transmission delay of feedback information from the receiver to the transmitter is not considered, the channels must be constant during, at least, $(M + N)$ STSs in our proposed AST, while they must be unchanged during at least $2N$ STSs in all conventional DSTBC techniques without the proposed AST. In practice, the difference between $(M + N)$ and $2N$ is usually very small. Hence, the proposed AST can be applied to various realistic cases where the coherence time of the channel is larger than $(M + N)$ STSs. In the case where the delay is considered, the channel coefficients must stay unchanged longer.*

The proposed general $(M, N; K)$ AST/DSTBC is more explicitly presented in Fig. 6.6. Steps (1a), (1b), (4) and (5) are carried out at the transmitter, while the remaining steps are carried out at the receiver. As stated earlier, step (1b) is not necessarily carried out after step (3a) finishes. Instead, the transmitter can perform step (1b) right after finishing step (1a). Similarly, because the matrix

$\mathbf{R}_0$ is straightforwardly created from the matrix $\tilde{\mathbf{R}}_0$, the receiver can perform step (3b) right after finishing step (3a). These properties reduce unnecessary delays during transmission.

6.3.5 The Restricted $(M, N; K)$ AST/DSTBC Scheme

As mentioned in (6.19), the number of feedback bits required in the *general* $(M, N; K)$ AST/DSTBC is

$$\mathcal{N} = \left\lceil \log_2 \binom{M}{N} \right\rceil$$

It is easy to realize that, $\mathcal{N}$ is large for large values of M and N. For instance, in the *general* (4,2;K) AST/DSTBC (K is arbitrary), we have $\mathcal{N} = 3$.

In the scenario where the capacity limitation of the feedback channel, especially in the uplink channels of the third generation (3G) mobile communication systems, needs to be considered, the number of feedback bits should be as small as possible. More importantly, limiting the number of feedback bits is necessary when fading is fast. Based on the *general* $(M, N; K)$ AST/DSTBC mentioned in Section 6.3.4, we modify the antenna selection criterion in (6.18) and propose here the so-called *restricted* $(M, N; K)$ AST/DSTBC for channels using DST-BCs. This modified AST is more amenable to practical implementation than the *general* $(M, N; K)$ AST/DSTBC since it provides a relatively good error performance, while requiring only one feedback bit for each channel coherence time T_c to inform the transmitter.

In the *restricted* $(M, N; K)$ AST/DSTBC, the set of M Tx antennas is divided into two subsets. Each subset includes N Tx antennas $(N < M)$. Subsets may partially overlap each other. Fig. 6.7 presents 3 cases for illustration. In Fig. 6.7(a), we give an example where 4 Tx antennas are divided into 2 subsets including 2 Tx antennas each, while in Fig. 6.7(b), 3 Tx antennas are divided into 2 subsets containing 2 Tx antennas each. These 2 cases can be applied, for instance, to the Alamouti DSTBC with the *restricted* (4,2;K) AST/DSTBC and with the *restricted* (3,2;K) AST/DSTBC, respectively. Fig. 6.7(c) shows another example where 5 Tx antennas are divided into 2 subsets which partially overlap one another and include 4 Tx antennas each. This case can be applied, for instance, to the order-4 DSTBC with the *restricted* (5,4;K) AST/DSTBC.

Let Ψ and Φ be the sets of indices indicating the orders of Tx antennas in each subset, respectively. The selection criterion of the *restricted* $(M, N; K)$ *AST/DSTBC* is as follows:

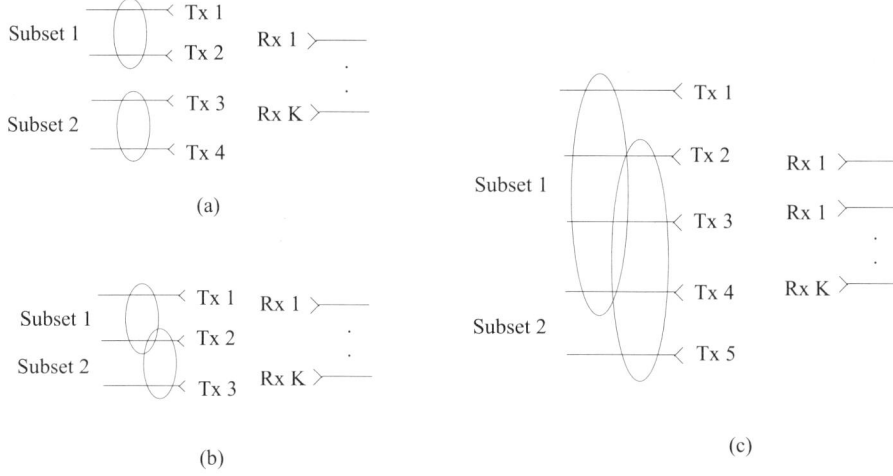

Figure 6.7. Some examples of the transmitter antenna grouping for (a) the *restricted* (4,2;K) AST/DSTBC, (b) the *restricted* (3,2;K) AST/DSTBC and (c) the *restricted* (5,4;K) AST/DSTBC.

During each coherence time T_c of the channel, the receiver compares

$$E_\Psi \stackrel{\Delta}{=} \sum_{j \in \Psi} \chi_j = \sum_{j \in \Psi} \left[\sum_{i=1}^{K} |r_{0ij}|^2 \right]$$

and

$$E_\Phi \stackrel{\Delta}{=} \sum_{j \in \Phi} \chi_j = \sum_{j \in \Phi} \left[\sum_{i=1}^{K} |r_{0ij}|^2 \right]$$

ie., the receiver compares the total power of the signals received by all K Rx antennas during the initial transmission from two subsets of Tx antennas, and then informs the transmitter to select the subset providing the greater total power. If E_Ψ is lager, then the transmitter selects the Tx antennas corresponding to the set of indices Ψ. Otherwise, the Tx antennas corresponding to the set of indices Φ should be selected. These procedures are repeated for different coherence durations T_c until the transmission of data is completed. It is obvious that only one feedback bit during each coherence time T_c is required for transmission diversity purpose.

6.3.6   Spatial Diversity Order of the Proposed ASTs

In this section, we consider the spatial diversity order of the *general* $(M, N; K)$ AST/DSTBC schemes proposed for channels using DSTBCs with *differential detection*. To do that, at first, we review the same issue for channels using STBCs with *coherent detection*. It is noted that, in this chapter, we consider the *transmitter* (*not receiver*) diversity antenna selection and the use of DSTBCs which have *orthogonal* structures. Therefore, it is useful to review the spatial diversity order of the ASTs associated with STBCs only (not STTCs or other STCs). Having this note in mind, we realize that the spatial diversity order of the ASTs for channels using STBCs has been intensively examined in the literature, such as in [Chen et al., 2003a], [Chen et al., 2003c], and [Gore and Paulraj, 2002].

Particularly, in [Chen et al., 2003a] and [Chen et al., 2003c], the authors limited themselves to consider the Alamouti STBC modulated by a binary phase shift keying (BPSK) signal constellation in the $(M, 2; 1)$ AST/STBC and $(M, 2; 2)$ AST/STBC schemes only. They proved there that these AST/STBC schemes provide a full spatial diversity order of M and $2M$, respectively, at high SNR ranges. A more exhaustive research is mentioned in [Gore and Paulraj, 2002]. In this paper, the authors proposed two antenna selection algorithms, i.e. ECK-based and SCK-based selection algorithms, for channels using STCs (such as STBCs or STTCs) with coherent detection. The authors proved there that, when ECK is available, i.e. the channel coefficients are exactly known at the receiver, antenna subset selection can provide a full spatial diversity order as if all antennas were used (see Section III in [Gore and Paulraj, 2002]). The authors also proved that, when SCK is available, i.e. the statistical channel knowledge is known at the receiver, antenna subset selection may provide a full spatial diversity order as if all antennas were used at high enough SNRs. In lower ranges of SNR, however, antenna subset selection may not provide full diversity (see Section IV B in [Gore and Paulraj, 2002]).

As opposed to space-time coded systems with *coherent* detection, for space-time coded systems with *non-coherent* detection, such as the systems using DSTBCs, the study on the spatial diversity order of AST/DSTBC schemes has not been examined yet. Due to this reason, in this chapter, we do not have ambition to provide an exhaustive study of this issue, which certainly requires further studies in the future.

Instead, we show that the problem of finding the spatial diversity order of the ASTs proposed for channels using DSTBCs with *differential detection* is similar to that for the case of STBCs with *coherent detection* when the SNR is high enough. Since the respective $(M, N; K)$ AST/STBC schemes in channels using STBCs can provide a full spatial diversity order [Chen et al., 2003a], [Chen et al., 2003c], [Gore and Paulraj, 2002], the *general* $(M, N; K)$ AST/DSTBC schemes in channels using DSTBCs can also provide a full spatial diversity order as if all antennas were used.

To facilitate our later analysis, we rewrite the Tx antenna selection criterion, which was mentioned in Section III A2 in [Gore and Paulraj, 2002], for channels using STBCs associated with ECK-based antenna selection. We use the superscript l ($l = 1, 2, 3, \ldots$) to indicate the different coherence durations of the channel. Since the ECK-based antenna selection is being considered, the channel coefficients between Tx and Rx antennas denoted by $\bar{a}_{ij}^{(l)}$, for $i = 1, \ldots, K$ and $j = 1, \ldots, M$, are perfectly known at the receiver. Let $\bar{\xi}_j^{(l)} = \sum_{i=1}^{K} |\bar{a}_{ij}^{(l)}|^2$. We assume that $\bar{a}_{ij}^{(l)}$s are i.i.d. complex Gaussian random variables with the distribution $\mathcal{CN}(0, \sigma_a)$.

With the notation mentioned in Section 6.3.2 of this chapter, we rewrite the Tx antenna selection criterion mentioned in Section III A2 in [Gore and Paulraj, 2002] for the $(M, N; K)$ AST/STBC scheme during the lth coherence duration as

$$
\begin{aligned}
\hat{\mathcal{I}}_N^{(l)} &= F_N\left(\bar{\xi}_1^{(l)}, \bar{\xi}_2^{(l)}, \ldots, \bar{\xi}_M^{(l)}\right) \\
&= F_N\left(\sum_{i=1}^{K} |\bar{a}_{i1}^{(l)}|^2, \sum_{i=1}^{K} |\bar{a}_{i2}^{(l)}|^2, \ldots, \sum_{i=1}^{K} |\bar{a}_{iM}^{(l)}|^2\right)
\end{aligned}
\tag{6.20}
$$

Now we return to consider our proposed, *general* $(M, N; K)$ AST/DSTBC for channels using DSTBCs with *differential detection*. The superscript k ($k = 1, 2, 3, \ldots, m$ where m is an arbitrary integer number) is used to indicate the different coherence durations of the considered channel (see Fig. 6.5). Since the differential detection is being considered, the channel coefficients between Tx and Rx antennas denoted by $a_{ij}^{(k)}$, for $i = 1, \ldots, K$; $j = 1, \ldots, M$; and $k = 1, \ldots, m$, are *unknown* at either the receiver or the transmitter.

As mentioned in Eq. (6.18) in Section 6.3.4, the selection criterion for the *general* $(M, N; K)$ AST/DSTBC during the kth coherence duration is

$$
\begin{aligned}
\hat{\mathcal{I}}_N^{(k)} & \\
&= F_N\left(\chi_1^{(k)}, \ldots, \chi_M^{(k)}\right) \\
&= F_N\left(\sum_{i=1}^{K} |a_{i1}^{(k)} + n_{0i1}^{(k)}|^2, \sum_{i=1}^{K} |a_{i2}^{(k)} + n_{0i2}^{(k)}|^2, \ldots, \sum_{i=1}^{K} |a_{iM}^{(k)} + n_{0iM}^{(k)}|^2\right)
\end{aligned}
\tag{6.21}
$$

We assume that the channel coefficients $a_{ij}^{(k)}$ and noise $n_{0ij}^{(k)}$ are i.i.d. complex Gaussian random variables with the distribution $\mathcal{CN}(0, \sigma_a)$ and $\mathcal{CN}(0, \sigma)$, respectively. We consider the mean and the variance of the following term:

$$\mu_{ij}^{(k)} \overset{\Delta}{=} |a_{ij}^{(k)} + n_{0ij}^{(k)}|^2$$

for $i = 1, \ldots, K$, $j = 1, \ldots, M$ and $k = 1, \ldots, m$.

Since $a_{ij}^{(k)}$ and $n_{0ij}^{(k)}$ are the i.i.d. zero-mean, complex Gaussian random variables, $(a_{ij}^{(k)} + n_{0ij}^{(k)})$ are the i.i.d., complex Gaussian random variables with the distribution $\mathcal{CN}(0, \rho)$ where $\rho = \sigma_a + \sigma$. Therefore, $\mu_{ij}^{(k)}$ are the i.i.d, *central* chi-squared random variables with $n = 2$ degrees of freedom and with the following mean and variance [Proakis, 2001] (pp. 42):

$$E\{\mu_{ij}^{(k)}\} = n\frac{\rho}{2} = \rho \, ,$$

$$\sigma_{\mu_{ij}^{(k)}} = 2n\left(\frac{\rho}{2}\right)^2 = \rho^2$$

We investigate the case in which the channel $SNR >> 1$. Equivalently, the variances of noise terms $n_{0ij}^{(k)}$s are very small in comparison with the variances of $a_{ij}^{(k)}$s, and therefore, $\rho \approx \sigma_a$. As a result, the means and the variances of $\mu_{ij}^{(k)}$ asymptotically approach

$$E\{\mu_{ij}^{(k)}\} \asymp \sigma_a \, ,$$

$$\sigma_{\mu_{ij}^{(k)}} \asymp \sigma_a^2 \tag{6.22}$$

when $SNR >> 1$. In (6.22), the sign $\asymp$ means the asymptotical approach.

On the other hand, we consider the following term:

$$\theta_{ij}^{(k)} = |a_{ij}^{(k)}|^2$$

for $i = 1, \ldots, K$, $j = 1, \ldots, M$ and $k = 1, \ldots, m$.

Similarly analyzed, $\theta_{ij}^{(k)}$ are the i.i.d, *central* chi-squared random variables having $n = 2$ degrees of freedom with the following mean and variance [Proakis, 2001] (pp. 42):

$$E\{\theta_{ij}^{(k)}\} = n\frac{\sigma_a}{2} = \sigma_a \, ,$$

$$\sigma_{\mu_{ij}^{(k)}} = 2n\left(\frac{\sigma_a}{2}\right)^2 = \sigma_a^2 \tag{6.23}$$

From (6.22) and (6.23), we realize that, $\mu_{ij}^{(k)}$s and $\theta_{ij}^{(k)}$s have the same statistical properties, i.e., means and variances when $SNR >> 1$. We can rewrite the antenna selection criterion of the *general* $(M, N; K)$ AST/DSTBC in (6.21) as

$$\hat{\mathcal{I}}_N^{(k)} \; \asymp \; \Gamma_N \left(\sum_{i=1}^K |a_{i1}^{(k)}|^2, \sum_{i=1}^K |a_{i2}^{(k)}|^2, \ldots, \sum_{i=1}^K |a_{iM}^{(k)}|^2 \right) \qquad (6.24)$$

when $SNR >> 1$.

Clearly, the antenna selection criterion for the *general* $(M, N; K)$ AST/DSTBC scheme now tends to be the same as the criterion mentioned in (6.20) for the $(M, N; K)$ AST/STBC scheme.

We may conclude that, if the channel $SNR \rightarrow \infty$, the *behaviour* of the *general* $(M, N; K)$ AST/DSTBC scheme proposed for channels using DST-BCs with *differential detection* tends to be the same as that of the $(M, N; K)$ AST/STBC scheme mentioned in the literature for channels using STBCs with *coherent detection*, although the *general* $(M, N; K)$ AST/DSTBC scheme is inferior by 3dB compared to the $(M, N; K)$ AST/STBC scheme due to the fact that the channel coefficients are not known at either transmitter or receiver. Since the $(M, N; K)$ AST/STBC schemes can achieve full spatial diversity as if all antennas were used [Chen et al., 2003a], [Chen et al., 2003c], [Gore and Paulraj, 2002], then so can the *general* $(M, N; K)$ AST/DSTBC schemes, provided that the channel SNR is high enough.

At lower ranges of the SNR, however, the antenna selection criterion (6.21) is ill-conditioned. In this scenario, antenna selection decisions are more likely erroneous. Hence, we can deduce that the *general* $(M, N; K)$ AST/DSTBC schemes may not provide full diversity when the SNR is small.

6.3.7   Simulation Results

In this section, we run some Monte-Carlo simulations to solidify our proposed AST/DSTBC schemes. We consider a wireless link comprising $K = 1$ Rx antenna. The channel SNR is defined to be the ratio between the total average power of the received signals and the average power of noise at the Rx antenna during each STS. In simulations, DSTBCs are modulated by a QPSK signal constellation.

First, the Alamouti DSTBC in (6.5) corresponding to N=2 is simulated. We consider the three following scenarios: 1) Alamouti DSTBC *without* ASTs; 2) Alamouti DSTBC with the *general* (3,2;1) AST/DSTBC (2 feedback bits); and 3) Alamouti DSTBC with the *restricted* (3,2;1) AST/DSTBC (1 feedback bit). Note that the number of feedback bits required for the *general* $(M, N; K)$ AST/DSTBC schemes is calculated by (6.19), while the number of feedback

Table 6.3. SNR gains (dB) of the general (3,2;1) AST/DSTBC and the restricted (3,2;1) AST/DSTBC in the channel using Alamouti DSTBC.

	General (3,2;1) AST/DSTBC	Restricted (3,2;1) AST/DSTBC
Feedback error 4%	3.25	2.9
Feedback error 10%	2.25	2.25

bit required for the *restricted* $(M, N; K)$ AST/DSTBC schemes is always 1. Furthermore, in each AST/DSTBC scheme, we examine two cases where the feedback error rates are assumed to be 4% and 10%. Transmitter antennas in the *restricted* (3,2;1) AST/DSTBC are grouped by the scheme mentioned in Fig. 6.7 (b).

Note that it would be better if we could compare the performances here with the performance of a DSTBC without ASTs which can provide the same spatial diversity order as the diversity order (equal to 3) provided by the *general* (3,2;1) AST/DSTBC scheme. This means that we should compare the performance of the Alamouti DSTBC (associated with the proposed ASTs) with that of an order-3 DSTBC (without ASTs). However, while the Alamouti DSTBC has a full rate, it is well known that DSTBCs of higher orders with a full rate do *not* exist. For this reason, it is unfair to compare the Alamouti DSTBC with an order-3 DSTBC because they have different code rates, and consequently, we do not plot the performance of any order-3 DSTBC in simulations.

As analyzed earlier, channel coefficients must be constant during at least two adjacent code blocks. If T_c denotes the coherence time of the channel, then it is required that:

- $T_c \geq 4$ STSs for the Alamouti DSTBC without ASTs, and

- $T_c \geq 5$ STSs for the Alamouti DSTBC with the *general* (3,2;1) AST/DSTBC or with the *restricted* (3,2;1) AST/DSTBC.

Therefore, to compare fairly the performance of the Alamouti DSTBC with different ASTs, the simulation is run for T_c which is *not less than 5 STSs*. Example 6.3.2 in Section 6.3.1.2 is one of such practical scenarios.

The performance of the Alamouti DSTBC with and without ASTs is shown in Fig. 6.8. Clearly, the proposed ASTs improve significantly the BER performance of the channel. The SNR gains (dB) in our proposed ASTs to achieve

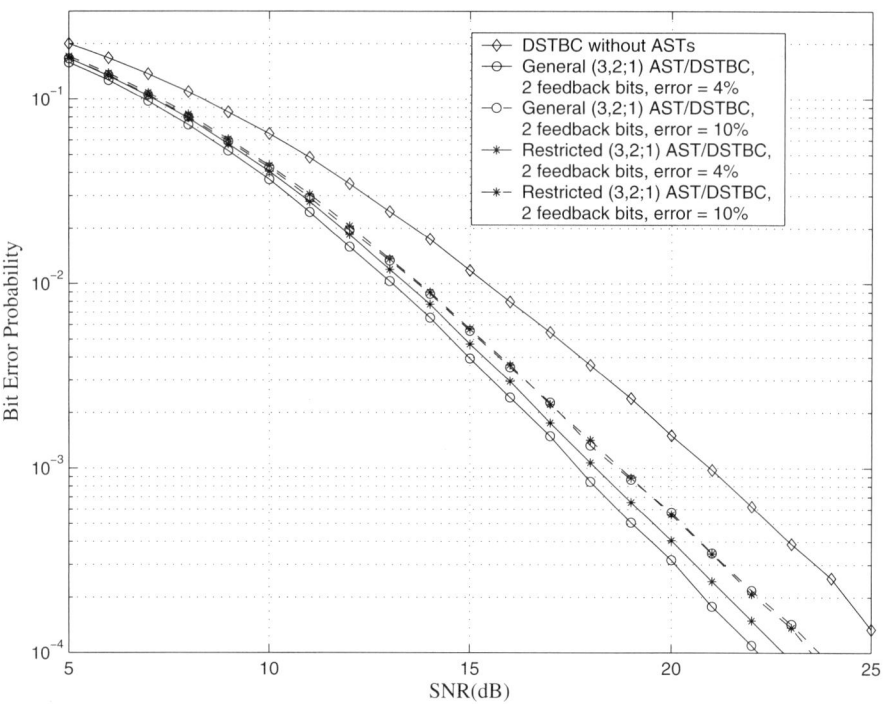

Figure 6.8. The Alamouti DSTBC with the *general* (3,2;1) AST/DSTBC and the *restricted* (3,2;1) AST/DSTBC schemes.

the same $BER = 10^{-3}$ as the Alamouti DSTBC without ASTs are given in Table 6.3.

Next, we examine the *square*, order-4, unitary DSTBC in (6.6) corresponding to $N = 4$ and the code rate 3/4. We consider the following three scenarios: 1) DSTBC without ASTs; 2) DSTBC with the *general* (5,4;1) AST/DSTBC (3 feedback bits); and 3) DSTBC with the *restricted* (5,4;1) AST/DSTBC (1 feedback bit). In each AST/DSTBC scheme, we also consider two cases where the feedback error rates are assumed to be 4% and 10%. Transmitter antennas in the *restricted* (5,4;1) AST/DSTBC are grouped by the scheme mentioned in Fig. 6.7 (c).

It is required that:

- $T_c \geq 8$ STSs for the considered DSTBC without ASTs, and

- $T_c \geq 9$ STSs for the considered DSTBC with the *general* (5,4;1) AST/DSTBC or with the *restricted* (5,4;1) AST/DSTBC.

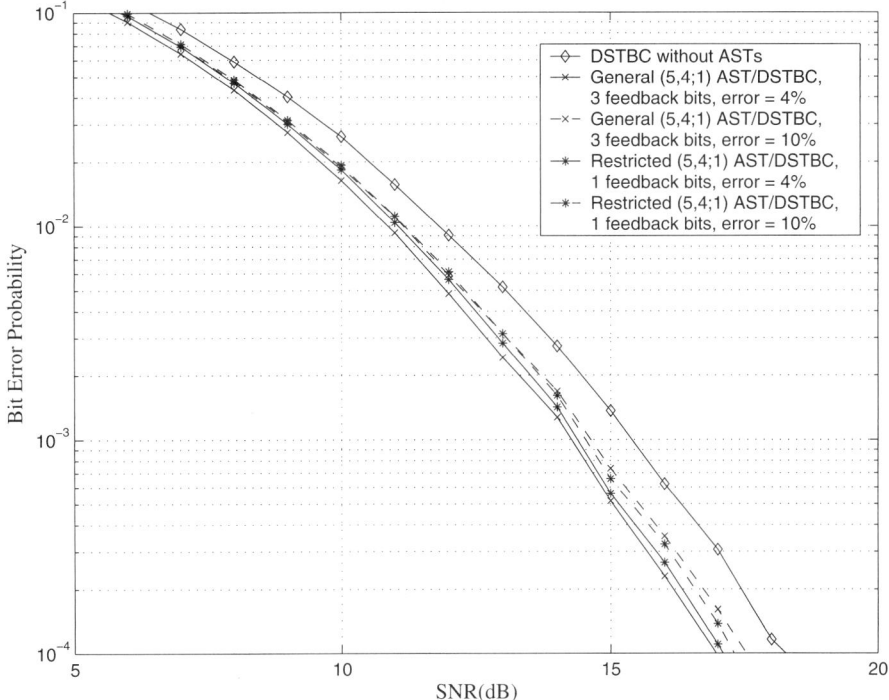

Figure 6.9. Square, order-4, unitary DSTBC with the *general* (5,4;1) AST/DSTBC and the *restricted* (5,4;1) AST/DSTBC schemes.

Table 6.4. SNR gains (dB) of the proposed (5,4;1) AST/DSTBC schemes in the channel using square, order-4, unitary DSTBC.

	General (5,4;1) AST/DSTBC	Restricted (5,4;1) AST/DSTBC
Feedback error 4%	1.2	1
Feedback error 10%	0.8	0.85

Therefore, the simulation is run for T_c which is *not less than 9 STSs*. Example 6.3.2 in Section 6.3.1.2 is still valid for this scenario.

The performance of the proposed AST/DSTBC schemes is presented in Fig 6.9. The SNR gains (dB) achieved by our proposed ASTs to have the same $BER = 10^{-3}$ as the DSTBC without ASTs are given in Table 6.4.

From all the aforementioned simulations, we realize that the proposed ASTs improve significantly the performance of wireless channels using DSTBCs. Also, we realize that the *restricted* $(M, N; K)$ AST/DSTBC schemes provide relatively good BER performance compared to the *general* $(M, N; K)$ AST/DSTBC schemes, while requiring only 1 feedback bit. More importantly, the *restricted* $(M, N; K)$ AST/DSTBC schemes may perform even better than the *general* $(M, N; K)$ AST/DSTBC schemes when the feedback error rate grows large (10% for instance). Intuitively, this is interpreted by the fact that the *restricted* AST/DSTBC schemes require only 1 feedback bit while the *general* AST/DSTBC schemes require multiple feedback bits. Therefore, when the feedback error rate grows large, the feedback information in the *restricted* ASTs is less likely erroneous than that in the *general* ASTs. As a result, the *restricted* AST/DSTBC schemes can be a good choice for the channels where fading changes fast.

6.4   Discussions and Conclusion

In this chapter, first, we have considered channels utilizing STBCs with *coherent detection*. Motivated by the N-out-of-M AST proposed in [Katz et al., 2001], we have proposed an improved AST referred to as the $(N + 1, N; K)$ AST/STBC scheme. The proposed AST provides the same BER performance as the N-out-of-$(N+1)$ AST proposed in [Katz et al., 2001], while shortening the time required to process feedback information. The time benefit is especially considerable for $N = 2$, such as, when the Alamouti STBC is utilized.

In the rest of the chapter, we have considered channels utilizing DSTBCs with *differential detection*. We have proposed two ASTs referred to as the *general* $(M, N; K)$ AST/DSTBC and the *restricted* $(M, N; K)$ AST/DSTBC schemes for the channels using DSTBCs with arbitrary number of Tx and Rx antennas.

Since the *general* $(M, N; K)$ AST/DSTBC scheme requires a large number of feedback bits when M, N and K are large, it is either impractical or uneconomical for implementation in such cases. Therefore, we have proposed the *restricted* $(M, N; K)$ AST/DSTBC schemes to overcome this shortcoming.

Particularly, the *restricted* $(M, N; K)$ AST/DSTBC is an attractive technique, which provides relatively good bit error performance, compared to the *general* $(M, N; K)$ AST/DSTBC, while requiring only one feedback bit. This advantage is very important in the case where the capacity limitation of the feedback channel, such as in the uplink channels of the 3G mobile communication systems, is considered. This advantage is also very beneficial in the channels where fading is fast and/or the feedback error rate in the feedback channel grows large.

Simulations show that the proposed ASTs improve noticeably the BER performance of wireless channels utilizing DSTBCs with a limited number (typically, 1 or 2) of training symbols per coherence time of the channel. The improvement is significant even in the case of 1 training symbol, i.e., in the *general* $(M, N; K)$ AST/DSTBC or *restricted* $(M, N; K)$ AST/DSTBC schemes where $M = (N + 1)$.

Further, the *restricted* $(M, N; K)$ AST/DSTBC schemes may provide better BER performance over the *general* $(M, N; K)$ AST/DSTBC schemes when the feedback error rate is large. Hence, the *restricted* AST/DSTBC schemes are a good choice for the channels where fading changes fast and/or the feedback error rate is large.

It is noted that, in this chapter, we have assumed that the carrier phase/frequency is perfectly recovered at the receiver. In fact, phase/frequency recovery errors may exist, which degrade the performance of the proposed ASTs. Those errors may occur due to the difference between the frequency of the local oscillators at the transmitter and the receiver, and/or due to the Doppler frequency shift effect. The effect of imperfect carrier recovery on the performance of the proposed ASTs in wireless channels utilizing DSTBCs will be examined in the next chapter.

Also, in this chapter, the delay of feedback information has not been considered. In reality, the delay of feedback information may somewhat degrade the overall performance of the proposed ASTs. Finally, as mentioned earlier, the *exhaustive* research on the spatial diversity order of the ASTs proposed for channel using DSTBCs has not been derived yet and it must be fully examined in the future work.

Chapter 7

PERFORMANCE OF DIVERSITY ANTENNA SELECTION TECHNIQUES IN IMPERFECT CHANNELS

7.1   Introduction

In Chapter 6, we have proposed two diversity antenna selection techniques (ASTs) for channels using Differential Space-Time Block Codes (DSTBCs), which are referred to as the AST/DSTBC schemes, in *uncorrelated* flat Rayleigh fading channels. Simulation results show that DSTBCs associated with the proposed ASTs provide much better bit error performance than those codes without ASTs.

In fact, transmission antennas including transmitter (Tx) antennas and receiver (Rx) antennas may not be spatially separated far enough from each other in order that the transmission coefficients of the channels between the Tx and Rx antennas are independent random variables. In other words, correlation between the transmission coefficients, which degrades the performance of the proposed ASTs, may be present. Therefore, this chapter analyzes the performance of our proposed AST/DSTBC schemes in such (spatially) correlated flat Rayleigh fading channels.

To do this, firstly, we propose here a very general, straightforward algorithm for generation of an *arbitrary* number of Rayleigh envelopes with any desired (*equal or unequal*) power in wireless channels either *with* or *without* Doppler frequency shift effects. The proposed algorithm can be applied to the case of either *spatial correlation*, such as with antenna arrays in MIMO systems, or *spectral correlation* between random processes like in Orthogonal Frequency Division Multiplexing (OFDM) systems. Besides being more generalized, the proposed algorithm is more accurate, while overcoming all shortcomings of the conventional methods.

Based on the proposed algorithm, the performance of our AST/DSTBC schemes proposed for systems utilizing DSTBCs is analyzed in spatially correlated Rayleigh fading channels. The analysis shows that our ASTs not only work well in uncorrelated Rayleigh fading channels, but also are very robust in correlated ones.

In addition, in Chapter 6, when proposing the AST/DSTBC schemes for channels using DSTBCs, we have also assumed that the carrier phase/frequency (or just phase, for short, except when it is clearly stated) is perfectly recovered at the receiver. In fact, phase recovery errors may exist, which degrade the performance of the proposed ASTs. Phase errors may occur due to the difference between the frequency of the local oscillators at the transmitter and at the receiver, and/or due to Doppler frequency-shift effects. Consequently, the effect of imperfect phase recovery on the performance of the proposed ASTs needs to be examined and proposing solutions to overcome the effect, if possible, is specially important.

This chapter thus examines the effect of imperfect carrier phase/frequency recovery at the receiver on the bit error performance of our diversity antenna selection techniques proposed for channels utilizing DSTBCs. The tolerance of DSTBCs associated with the proposed ASTs to phase/frequency errors is then analyzed. The content of this chapter is based on our following papers [Tran et al., 2005a], [Tran et al., 2005b], [Tran et al., 2005c], [Tran et al., 2004c].

The chapter is organized as follows. In Section 7.2, we propose the generalized algorithm for the generation of correlated Rayleigh fading envelopes. Section 7.3 analyzes the performance of our AST/DSTBC schemes proposed for channels using DSTBCs in spatially correlated, flat Rayleigh fading scenario. Section 7.4 analyzes the effect of imperfect carrier frequency/phase recovery on the performance of our proposed AST/DSTBC schemes in uncorrelated, flat Rayleigh fading channels. The chapter is concluded by Section 7.5.

7.2   A Generalized Algorithm for the Generation of Correlated Rayleigh Fading Envelopes

7.2.1   Shortcomings of Conventional Methods

Generation of correlated Rayleigh fading envelopes has been intensively examined in the literature, such as [Beaulieu, 1999], [Beaulieu and Merani, 2000], [de Leon et al., 2004a], [de Leon et al., 2004b], [Ertel and Reed, 1998], [Hansson and Aulin, 1999], [Natarajan et al., 2000], [Salz and Winters, 1994], [Sorooshyari and Daut, 2003], [Verdin and Tozer, 1993], [Young and Beaulieu, 1996], [Young and Beaulieu, 1998], and [Young and Beaulieu, 2000]. However, all conventional methods have their own shortcomings, which seriously limit their applicability. To point this out more clearly, we first analyze the shortcomings of some conventional methods.

In [Salz and Winters, 1994], the authors derived fading correlation properties in antenna arrays and, then, briefly mentioned the algorithm to generate complex Gaussian random variables (with Rayleigh envelopes) corresponding to a desired correlation coefficient matrix. This algorithm was proposed for generating *equal power* Rayleigh envelopes only, rather than *arbitrary* (*equal or unequal*) power Rayleigh envelopes.

In the papers [Beaulieu, 1999] and [Ertel and Reed, 1998], the authors proposed different methods for generating only $N=2$ *equal power* correlated Rayleigh envelopes. In [Beaulieu and Merani, 2000], the authors generalized the method in [Beaulieu, 1999] for $N \geq 2$. However, in this method, Cholesky decomposition [Adeli and Soegiarso, 1999] is used, and consequently, the covariance matrix must be positive definite, which is not always realistic. An example, where the covariance matrix is *not* positive definite, is derived later in Example 7.2.1 of Section 7.2.4.1 of this chapter.

These methods were then more generalized in [Natarajan et al., 2000], where one can generate *any number* of Rayleigh envelopes corresponding to a desired covariance matrix and with *any power*, i.e., even with *unequal* power. However, again, the covariance matrix *must be positive definite* in order for Cholesky decomposition to be performable. In addition, the authors in [Natarajan et al., 2000] forced the covariances of the complex Gaussian random variables (with Rayleigh fading envelopes) to be *real* (see Eq. (8) in [Natarajan et al., 2000]). This limitation is *incorrect* in various cases because, in fact, the covariances of the complex Gaussian random variables are more likely to be complex.

In [Sorooshyari and Daut, 2003], the authors proposed a method for generating *any number* of Rayleigh envelopes with *equal power* only. Although the method in [Sorooshyari and Daut, 2003] works well in various cases, it however *fails* to perform a Cholesky decomposition for some complex covariance matrices in MatLab due to the roundoff errors of Matlab [1]. This shortcoming is overcome by some modifications mentioned later in our proposed algorithm.

---

[1] It has been well known that Cholesky decomposition may not work for the matrix having eigenvalues being equal or close to zeros. We consider the following covariance matrix $\mathcal{K}$, for instance:

$$\mathcal{K} = \begin{bmatrix} 1.04361 & 0.7596 - 0.3840i & 0.6082 - 0.4427i & 0.4085 - 0.8547i \\ 0.7596 + 0.3840i & 1.04361 & 0.7780 - 0.3654i & 0.6082 - 0.4427i \\ 0.6082 + 0.4427i & 0.7780 + 0.3654i & 1.04361 & 0.7596 - 0.3840i \\ 0.4085 + 0.8547i & 0.6082 + 0.4427i & 0.7596 + 0.3840i & 1.04361 \end{bmatrix}$$

Cholesky decomposition does not work for this covariance matrix although it is positive definite.

More importantly, the method proposed in [Sorooshyari and Daut, 2003] *fails* to generate Rayleigh fading envelopes corresponding to a desired covariance matrix in a real-time scenario *where Doppler frequency shifts are considered*. This is because passing Gaussian random variables with variances assumed to be unit (for simplicity of explanation) through a Doppler filter changes remarkably the variances of those variables. The variances of the variables at the outputs of Doppler filters are *not* unit any more, but depend on the variances of the variables at the inputs of the filters as well as the characteristics of those filters. The authors in [Sorooshyari and Daut, 2003] did not realize this variance-changing effect caused by Doppler filters. We will return to this issue later in this chapter.

For the aforementioned reasons, a *more generalized* algorithm is required to generate *any number* of Rayleigh fading envelopes with *any power* (*equal or unequal* power) corresponding to *any desired* covariance matrix. The algorithm should be applicable to both *discrete time instant* scenario and *real-time* scenario. The algorithm is also expected to overcome roundoff errors which may cause the interruption of MatLab programs. In addition, the algorithm should work well, regardless of the positive definiteness of the covariance matrices. Furthermore, the algorithm should provide a straightforward method for the generation of complex Gaussian random variables (with Rayleigh envelopes) with correlation properties as functions of *time delay* and *frequency separation* (such as in OFDM systems), or *spatial separation* between transmission antennas (like with antenna arrays in MIMO systems). This chapter proposes such an algorithm, which can also be found in our papers [Tran et al., 2005a], [Tran et al., 2005b], [Tran et al., 2005c].

7.2.2   Fading Correlation as Functions of Time Delay and Frequency Separation

In [Jakes, 1974], Jakes considered the scenario where all complex Gaussian random processes with Rayleigh envelopes have *equal* power σ^2 and derived the correlation properties between random processes as functions of both time delay and frequency separation, such as in OFDM systems. Let $z_k(t)$ and $z_j(t)$ be the two zero-mean complex Gaussian random processes at time instant t, corresponding to frequencies f_k and f_j, respectively. Denote

$$x_k \overset{\Delta}{=} Re(z_k(t)); \qquad y_k \overset{\Delta}{=} Im(z_k(t));$$
$$x_j \overset{\Delta}{=} Re(z_j(t + \tau_{k,j})); \quad y_j \overset{\Delta}{=} Im(z_j(t + \tau_{k,j}))$$

where $\tau_{k,j}$ is the arrival time delay between two signals and $Re(.)$, $Im(.)$ are the real and imaginary parts of the argument, respectively. By definition, the covariances between the real and imaginary parts of $z_k(t)$ and $z_j(t + \tau_{k,j})$ are

$$R_{xx\,k,j} \overset{\Delta}{=} E(x_k x_j); \quad R_{yy\,k,j} \overset{\Delta}{=} E(y_k y_j); \tag{7.1}$$

$$R_{xy\,k,j} \overset{\Delta}{=} E(x_k y_j); \quad R_{yx\,k,j} \overset{\Delta}{=} E(y_k x_j) \tag{7.2}$$

Those covariances have been derived in [Jakes, 1974] (see Eq. (1.5-20)) as

$$R_{xx\,k,j} = R_{yy\,k,j} = \frac{\sigma^2 J_0(2\pi F_m \tau_{k,j})}{2[1 + (\Delta\omega_{k,j}\sigma_\tau)^2]} ; \tag{7.3}$$

$$R_{xy\,k,j} = -R_{yx\,k,j} = -\Delta\omega_{k,j}\sigma_\tau R_{xx\,k,j} \tag{7.4}$$

where

- σ^2 is the variance (power) of the complex Gaussian random processes ($\frac{\sigma^2}{2}$ is the variance per dimension);

- J_0 is the first-kind Bessel function of the zeroth-order;

- F_m is the maximum Doppler frequency $F_m = \frac{v}{\lambda} = \frac{vf_c}{c}$. In this formula, λ is the wavelength of the carrier, f_c is the carrier frequency, c is the speed of light and v is the mobile speed;

- $\Delta\omega_{k,j} = 2\pi(f_k - f_j)$ is the angular frequency separation between the two complex Gaussian processes with Rayleigh envelopes at frequencies f_k and f_j;

- σ_τ is the root-mean-square (rms) delay spread of the wireless channel.

It should be emphasized that, the equalities (7.3) and (7.4) hold only when the set of *multi-path channel coefficients*, which were denoted as C_{nm} and derived in Eq. (1.5-1) and (1.5-2) in [Jakes, 1974], as well as the *power* are assumed to be the *same* for different random processes (with different frequencies). Readers may refer to [Jakes, 1974] (pp. 46–49) for the explicit exposition.

7.2.3   Fading Correlation as Functions of Spatial Separation in Antenna Arrays

Fading correlation properties between wireless channels as functions of antenna spacing in antenna arrays have been mentioned in [Salz and Winters, 1994]. Fig. 7.1 presents a typical model of the channel where all signals

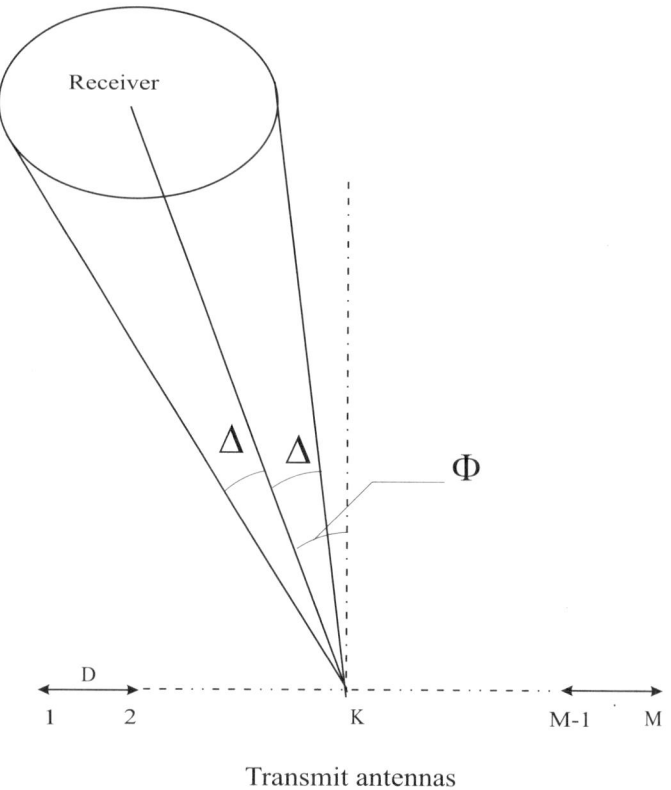

Figure 7.1.   Model to examine the spatial correlation between transmitter antennas.

from a receiver are assumed to arrive at M Tx antennas within $\pm\Delta$ at angle Φ ($|\Phi| \leq \pi$). Let λ be the wavelength, D the distance between the two adjacent Tx antennas, and $z = 2\pi\frac{D}{\lambda}$. In [Salz and Winters, 1994], it is assumed that fading corresponding to different receivers is independent. This is reasonable if receivers are not on top of each other within some wavelengths and they are surrounded by their own scatterers. Consequently, we only need to calculate the correlation properties for a typical receiver. The fading in the channel between a given k^{th} Tx antenna and the receiver may be considered as a zero-mean, complex Gaussian random variable, which is presented as $b^{(k)} = x^{(k)} + iy^{(k)}$. We denote the covariances between the real parts and between the imaginary parts of the fading corresponding to the kth and j^{th} Tx antennas[2] to be $R_{xxk,j}$ and $R_{yyk,j}$, respectively, while those terms between the real and imaginary parts

---

[2]Note that k and j here are antenna indices, while they are frequency indices in Section 7.2.2.

of the fading to be $R_{xyk,j}$ and $R_{yxk,j}$. The terms $R_{xxk,j}$, $R_{yyk,j}$, $R_{xyk,j}$ and $R_{yxk,j}$ are similarly defined as (7.1) and (7.2). Then, it has been proved that the closed-form expressions of these covariances normalized by the variance per dimension (real and imaginary) are (see Eq. (A. 19) and (A. 20) in [Salz and Winters, 1994])

$$
\begin{aligned}
\tilde{R}_{xxk,j} &= \tilde{R}_{yyk,j} \\
&= J_0(z(k-j)) + 2\sum_{m=1}^{\infty} J_{2m}(z(k-j))cos(2m\Phi)\frac{sin(2m\Delta)}{2m\Delta}
\end{aligned}
$$

$$(7.5)$$

$$
\begin{aligned}
\tilde{R}_{xyk,j} &= -\tilde{R}_{yxk,j} \\
&= 2\sum_{m=0}^{\infty}\left[J_{2m+1}(z(k-j))sin[(2m+1)\Phi]\frac{sin[(2m+1)\Delta]}{(2m+1)\Delta}\right]
\end{aligned}
$$

$$(7.6)$$

where

$$
\tilde{R}_{k,j} = \frac{2R_{k,j}}{\sigma^2}
$$

In other words, we have

$$
R_{k,j} = \frac{\sigma^2 \tilde{R}_{k,j}}{2} \tag{7.7}
$$

In these equations, J_q is the first-kind Bessel function of the integer order q, and $\sigma^2/2$ is the variance per dimension of the received signal at each Tx antenna, i.e., it is assumed in [Salz and Winters, 1994] that signals corresponding to different Tx antennas have *equal* variances σ^2.

Similarly to Section 7.2.2, the equalities (7.5) and (7.6) hold only when the set of *multi-path channel coefficients*, which were denoted as g_n and derived in Eq. (A-1) in [Salz and Winters, 1994], and the *power* are assumed to be the *same* for different random processes. Readers may refer to [Salz and Winters, 1994] (pp. 1054–1056) for the explicit exposition.

### 7.2.4	Generalized Algorithm to Generate Correlated, Flat Rayleigh Fading Envelopes

#### 7.2.4.1	Covariance Matrix of Complex Gaussian Random Variables with Rayleigh Fading Envelopes

It is known that Rayleigh fading envelopes can be generated from zero-mean, complex Gaussian random variables. We consider here a column vector $\mathbf{z}$ of N zero-mean, complex Gaussian random variables with variances (or powers) $\sigma_{g_j}^2$, for $j = 1, \ldots, N$. Denote $\mathbf{z} = (z_1, \ldots, z_N)^T$, where z_j $(j = 1, \ldots, N)$ is regarded as

$$z_j = r_j e^{i\theta_j} = x_j + iy_j$$

The modulus of z_j is

$$r_j = \sqrt{x_j^2 + y_j^2}$$

It is assumed that the phases θ_js are independent, identically uniformly distributed random variables. As a result, the real and imaginary parts of each z_j are independent (but z_js are not necessarily independent), i.e., the covariances $E(x_j y_j) = 0$ for $\forall j$ and therefore, r_js are *Rayleigh envelopes*.

Let $\sigma_{g_{xj}}^2$ and $\sigma_{g_{yj}}^2$ be the variances per dimension (real and imaginary), i.e., $\sigma_{g_{xj}}^2 = E(x_j^2)$, $\sigma_{g_{yj}}^2 = E(y_j^2)$. Clearly, $\sigma_{g_j}^2 = \sigma_{g_{xj}}^2 + \sigma_{g_{yj}}^2$. If $\sigma_{g_{xj}}^2 = \sigma_{g_{yj}}^2$, then $\sigma_{g_{xj}}^2 = \sigma_{g_{yj}}^2 = \frac{\sigma_{g_j}^2}{2}$. Note that we consider a very general scenario where the variances (powers) of the real parts are not necessarily equal to those of the imaginary parts. Also, the powers of Rayleigh envelopes denoted as $\sigma_{r_j}^2$ are not necessarily equal to one another. Therefore, the scenario where the variances of Rayleigh envelopes are equal to one another, or the variances of real parts are equal to those of imaginary parts, such as the scenario mentioned in either Section 7.2.2 or Section 7.2.3, is considered as a particular case.

For $k \neq j$, we define the covariances $R_{xx_{k,j}}$, $R_{yy_{k,j}}$, $R_{xy_{k,j}}$ and $R_{yx_{k,j}}$ between the real as well as imaginary parts of z_k and z_j, similarly to those mentioned in (7.1) and (7.2).

By definition, the covariance matrix $\mathcal{K}$ of $\mathbf{z}$ is

$$\mathcal{K} = E(\mathbf{z}\mathbf{z}^H) \triangleq [\mu_{k,j}]_{N \times N} \tag{7.8}$$

where $(.)^H$ denotes the Hermitian transposition operation and

$$\mu_{k,j} = \begin{cases} \sigma_{g_j}^2 & \text{if } k \equiv j \\ (R_{xx_{k,j}} + R_{yy_{k,j}}) - i(R_{xy_{k,j}} - R_{yx_{k,j}}) & \text{if } k \neq j \end{cases} \tag{7.9}$$

In reality, the estimated covariance matrix $\mathcal{K}$ is *not* always positive semi-definite. An example where the covariance matrix $\mathcal{K}$ is *not* positive semi-definite is derived as follows.

EXAMPLE 7.2.1 *We examine an antenna array comprising 3 Tx antennas. Let D_{kj}, for $k, j = 1, \ldots, 3$, be the distance between the kth and j^{th} antennas. The distance D_{jk} between the jth and k^{th} antennas is then $D_{jk} = -D_{kj}$. Specifically, we consider the case*

$$
\begin{aligned}
D_{21} &= 0.0385\lambda \\
D_{31} &= 0.1789\lambda \\
D_{32} &= 0.1560\lambda
\end{aligned}
$$

where λ is the wavelength. Clearly, these antennas are neither equally spaced, nor positioned in a straight line. Instead, they are positioned at the 3 peaks of a triangle.

If the Rx antenna is far enough from the Tx antennas, we can assume that all signals from the receiver arrive at the Tx antennas within $\pm\Delta$ at angle Φ (see Fig. 7.1 for the illustration of these notations). As a result, the analytical results mentioned in Section 7.2.3 with small modifications can still be applied to this case. In particular, the covariance matrix $\mathcal{K}$ can still be calculated following (7.5), (7.6), (7.7), (7.8) and (7.9), provided that, in (7.5) and (7.6), the products $z(k - j)$ (or $2\pi D(k - j)/\lambda$) are replaced by $2\pi D_{kj}/\lambda$. This is because, in our considered case, D_{kj} are the actual distances between the kth and j^{th} Tx antennas, for $k, j = 1, \ldots, 3$.

Further, we assume that the variance σ^2 of the received signals at each Tx antenna in (7.7) is unit, i.e., $\sigma^2 = 1$. We also assume that $\Phi = 0.1114\pi$ rad and $\Delta = 0.1114\pi$ rad.

In order to examine the performance of the considered system, Rayleigh fading envelopes are required to be simulated. In turn, the covariance matrix of the complex Gaussian random variables corresponding to these Rayleigh envelopes must be calculated. Based on the aforementioned assumptions, from the theoretically analytical equations (7.5), (7.6), (7.7), and the definition equations (7.8) and (7.9), we have the following desired covariance matrix for the considered configuration of Tx antennas:

$$
\mathcal{K} = \begin{bmatrix}
1.0000 & 0.9957 + 0.0811i & 0.9090 + 0.3607i \\
0.9957 - 0.0811i & 1.0000 & 0.9303 + 0.3180i \\
0.9090 - 0.3607i & 0.9303 - 0.3180i & 1.0000
\end{bmatrix}
$$

$$(7.10)$$

Performing an eigen decomposition, we have the following eigenvalues -
0.0092; 0.0360; and 2.9733. Therefore, $\mathcal{K}$ is not positive semi-definite. This
also means that $\mathcal{K}$ is not positive definite.

It is important to emphasize that, from the mathematical point of view, covariance matrices are *always* positive semi-definite by definition (7.8), i.e. the
eigenvalues of the covariance matrices are *either* zero *or* positive. However,
this does not contradict the above example where the covariance matrix $\mathcal{K}$ has
a negative eigenvalue. The main reason why the desired covariance matrix $\mathcal{K}$
is not positive semi-definite is due to the approximation and the simplifications
of the model mentioned in Fig. 7.1 in calculating the covariance values, i.e.,
due to the preciseness of the equations (7.5) and (7.6), compared to the true
covariance values. In other words, errors in estimating covariance values may
exist in the calculation. Those errors may result in a covariance matrix being
not positive semi-definite.

A question that could be raised here is why the covariance matrix of *complex*
Gaussian random variables (with Rayleigh fading envelopes), rather than the
covariance matrix of *Rayleigh envelopes*, is of particular interest. This is due
to the two following reasons.

From the physical point of view, in the covariance matrix of *Rayleigh en-*
velopes, the correlation properties R_{xx}, R_{yy} of the real components (in-phase
components) as well as the imaginary components (quadrature phase components) themselves and the correlation properties R_{xy}, R_{yx} between the real
and imaginary components of random variables are *not* directly present (these
correlation properties are defined in (7.1) and (7.2)). On the contrary, those
correlation properties are *clearly* present in the covariance matrix of complex
Gaussian random variables with the desired Rayleigh envelopes. In other words,
the covariance matrix of complex Gaussian random variables with the desired
Rayleigh envelopes presents the physical significance of the correlation properties of random variables in more detail than the covariance matrix of *Rayleigh*
envelopes.

Further, from the mathematical point of view, it is possible to have an one-to-
one mapping *from* the cross-correlation coefficients ρ_{gij} (between the ith and
j^{th} complex Gaussian random variables) *to* the cross-correlation coefficients
ρ_{rij} (between Rayleigh fading envelopes) as follows (see Eq. (1.5-26) in [Jakes,
1974]):

$$\rho_{rij} = \frac{(1 + |\rho_{gij}|)E_{int}\left(\frac{2\sqrt{|\rho_{gij}|}}{1+|\rho_{gij}|}\right) - \frac{\pi}{2}}{2 - \frac{\pi}{2}}$$

where $E_{int}(.)$ is the complete elliptic integral of the second kind. Some good approximations of this relationship between ρ_{rij} and ρ_{gij} are presented in the mapping Table II in [Ertel and Reed, 1998], the look-up Table I in [Natarajan et al., 2000] and Fig. 1 in [Natarajan et al., 2000].

However, the reverse mapping, i.e. the mapping *from* ρ_{rij} *to* ρ_{gij}, is *multivalent*. It means that, for a given ρ_{rij}, we have to determine ρ_{gij} somehow in order to generate Rayleigh fading envelopes and the possible values of ρ_{gij} may be significantly different from each other depending on how ρ_{gij} is determined from ρ_{rij}. It is noted that ρ_{rij} is always real, but ρ_{gij} may be complex.

For the two aforementioned reasons, the covariance matrix of complex Gaussian random variables (with Rayleigh envelopes), as opposed to the covariance matrix of Rayleigh envelopes, is of particular interest in this chapter.

7.2.4.2 Forced Positive Semi-definiteness of the Covariance Matrix

First, we need to define the *coloring matrix* $\mathcal{L}$ corresponding to a covariance matrix $\mathcal{K}$. The *coloring matrix* $\mathcal{L}$ is defined to be the $N \times N$-sized matrix satisfying

$$\mathcal{L}\mathcal{L}^H = \mathcal{K}$$

It is noted that the coloring matrix is *not* necessarily a lower triangular matrix. Particularly, to determine the coloring matrix $\mathcal{L}$ corresponding to a covariance matrix $\mathcal{K}$, we can use *either* Cholesky decomposition [Adeli and Soegiarso, 1999] as mentioned in a number of papers, which have been reviewed in Section 7.2.1 of this chapter, *or* eigen decomposition which is mentioned in the next section of this chapter. The former yields a lower triangular coloring matrix, while the later yields a square coloring matrix.

Unlike Cholesky decomposition, where the covariance matrix $\mathcal{K}$ must be *positive definite*, eigen decomposition requires that $\mathcal{K}$ is at least *positive semidefinite*, i.e. the eigenvalues of $\mathcal{K}$ are either zeros or positive. We will explain later why the covariance matrix must be positive semi-definite even in the case where eigen decomposition is used to calculate the coloring matrix. The covariance matrix $\mathcal{K}$, in fact, may *not* be positive semi-definite, i.e. $\mathcal{K}$ may have negative eigenvalues, as the case mentioned in Example 7.2.1 of Section 7.2.4.1.

To overcome this obstacle, similarly to (but not exactly as) the method in [Sorooshyari and Daut, 2003], we approximate the covariance matrix being *not* positive semi-definite by a positive semi-definite one. This procedure is presented as follows.

Assume that $\mathcal{K}$ is the desired covariance matrix, which is *not* positive semi-definite. Perform the *eigen decomposition* $\mathcal{K} = \mathbf{V}\mathbf{G}\mathbf{V}^H$, where $\mathbf{V}$ is the matrix of eigenvectors and $\mathbf{G}$ is a diagonal matrix of eigenvalues of the matrix $\mathcal{K}$. Let $\mathbf{G} = diag(\lambda_1, \ldots, \lambda_N)$. Calculate the approximate matrix $\mathbf{\Lambda} \triangleq diag(\hat{\lambda}_1, \ldots, \hat{\lambda}_N)$, where

$$\hat{\lambda}_j = \begin{cases} \lambda_j & \text{if } \lambda_j \geq 0 \\ 0 & \text{if } \lambda_j < 0 \end{cases}$$

We now compare our approximation procedure to the approximation procedure mentioned in [Sorooshyari and Daut, 2003]. The authors in [Sorooshyari and Daut, 2003] used the following approximation:

$$\hat{\lambda}_j = \begin{cases} \lambda_j & \text{if } \lambda_j > 0 \\ \varepsilon & \text{if } \lambda_j \leq 0 \end{cases}$$

where ε is a small, positive, real number.

Clearly, besides overcoming the disadvantage of Cholesky decomposition, our approximation procedure is *more precise* than the one mentioned in [Sorooshyari and Daut, 2003] since the matrix $\mathbf{\Lambda}$ in our algorithm approximates to the matrix $\mathbf{G}$ better than the one mentioned in [Sorooshyari and Daut, 2003]. Therefore, the desired covariance matrix $\mathcal{K}$ is well approximated by the positive semi-definite matrix $\mathbf{K} = \mathbf{V}\mathbf{\Lambda}\mathbf{V}^H$ from the Frobenius point of view [Sorooshyari and Daut, 2003].

7.2.4.3   Determining the Coloring Matrix Using Eigen Decomposition

In most of the conventional methods, a Cholesky decomposition was used to determine the coloring matrix. As analyzed earlier in Section 7.2.1, Cholesky decomposition may not work for the covariance matrix which has eigenvalues being equal or close to zeros.

To overcome this disadvantage, we use eigen decomposition, instead of Cholesky decomposition, to calculate the coloring matrix. A comparison of the computational efforts between the two methods (eigen decomposition versus Cholesky decomposition) is mentioned later in this chapter. The coloring matrix is calculated as follows:

At this stage, we have the forced positive semi-definite covariance matrix $\mathbf{K}$, which is equal to the desired covariance matrix $\mathcal{K}$ if $\mathcal{K}$ is positive semi-definite, or is an approximate matrix of $\mathcal{K}$ otherwise. Further, as mentioned earlier, we have $\mathbf{K} = \mathbf{V}\mathbf{\Lambda}\mathbf{V}^H$, where $\mathbf{\Lambda} = diag(\hat{\lambda}_1, \ldots, \hat{\lambda}_N)$ is the matrix of eigenvalues of $\mathbf{K}$. Since $\mathbf{K}$ is a positive semi-definite matrix, it follows that $\{\hat{\lambda}_j\}_{j=1}^N$ are *real* and *non-negative*.

We now calculate a new matrix $\bar{\mathbf{\Lambda}}$ as

$$\bar{\mathbf{\Lambda}} \;=\; \sqrt{\mathbf{\Lambda}} = diag\left(\sqrt{\hat{\lambda}_1}, \ldots, \sqrt{\hat{\lambda}_N}\right) \tag{7.11}$$

Clearly, $\bar{\mathbf{\Lambda}}$ is a *real, diagonal* matrix that results in

$$\bar{\mathbf{\Lambda}}\bar{\mathbf{\Lambda}}^H \;=\; \bar{\mathbf{\Lambda}}\bar{\mathbf{\Lambda}} = \mathbf{\Lambda} \tag{7.12}$$

If we denote $\mathbf{L} \overset{\Delta}{=} \mathbf{V}\bar{\mathbf{\Lambda}}$, then it follows that

$$\mathbf{L}\mathbf{L}^H \;=\; (\mathbf{V}\bar{\mathbf{\Lambda}})(\mathbf{V}\bar{\mathbf{\Lambda}})^H = \mathbf{V}\bar{\mathbf{\Lambda}}\bar{\mathbf{\Lambda}}^H\mathbf{V}^H = \mathbf{V}\mathbf{\Lambda}\mathbf{V}^H = \mathbf{K} \tag{7.13}$$

It means that the coloring matrix $\mathbf{L}$ corresponding to the covariance matrix $\mathbf{K}$ can be computed *without* using Cholesky decomposition. Thereby, the shortcoming of the paper [Sorooshyari and Daut, 2003], which is related to round-off errors in Matlab caused by Cholesky decomposition and is pointed out in Section 7.2.1, can be overcome.

We now explain why the covariance matrix must be positive semi-definite even when an eigen decomposition is used to compute the coloring matrix. It is easy to realize that, if $\mathbf{K}$ is *not* positive semi-definite, then $\bar{\mathbf{\Lambda}}$ calculated by (7.11) is a *complex* matrix. As a result, (7.12) and (7.13) are not satisfied.

7.2.4.4   Proposed Algorithm

In Section 7.2.1, we have shown that the method proposed in [Sorooshyari and Daut, 2003] fails to generate Rayleigh fading envelopes corresponding to a desired covariance matrix in a real-time scenario where Doppler frequency shifts are considered. This is because the authors in [Sorooshyari and Daut, 2003] did not consider the variance-changing effect caused by Doppler filters.

To surmount this shortcoming, the two following *simple, but important* modifications must be carried out:

1 Unlike step 6 of the method in [Sorooshyari and Daut, 2003], where N independent, complex Gaussian random variables (with Rayleigh fading envelopes) are generated with *unit* variances, in our algorithm, this step is modified in order to be able to generate independent, complex Gaussian random variables with *arbitrary* variances σ_g^2. Correspondingly, step 7 of the method in [Sorooshyari and Daut, 2003] must also be modified. Besides being more generalized, the modification of our algorithm in steps 6 and 7 allows us to combine correctly the outputs of Doppler filters in the method proposed in [Young and Beaulieu, 2000] and our algorithm.

2 The variance-changing effect of Doppler filters must be considered. It means that, we have to calculate the variance of the outputs of Doppler filters, which may have an *arbitrary* value depending on the variance of the complex Gaussian random variables at the inputs of Doppler filters as well as the characteristics of those filters. The variance value of the outputs is then input into the step 6 which has been modified as mentioned above.

The modification 1 can be carried out in the algorithm generating Rayleigh fading envelopes in a *discrete-time* scenario (see the algorithm mentioned in this section). The modification 2 can be carried out in the algorithm generating Rayleigh fading envelopes in a real-time scenario *where Doppler frequency shifts are considered* (see the algorithm mentioned in Section 7.2.5).

From the above observations, we propose here a generalized algorithm to generate N correlated Rayleigh envelopes in a *single time instant* as:

1 In a general case, the desired variances (powers) $\{\sigma_{g_j}^2\}_{j=1}^N$ of complex Gaussian random variables with Rayleigh envelopes must be known. Specially, if the desired variances (powers) $\{\sigma_{r_j}^2\}_{j=1}^N$ of Rayleigh envelopes are known, then $\{\sigma_{g_j}^2\}_{j=1}^N$ are calculated as follows [3]:

$$\sigma_{g_j}^2 = \frac{\sigma_{r_j}^2}{\left(1 - \frac{\pi}{4}\right)} \quad \forall j = 1 \ldots N \tag{7.14}$$

2 From the desired correlation properties of correlated complex Gaussian random variables with Rayleigh envelopes, determine the covariances $R_{xx k,j}$, $R_{yy k,j}$, $R_{xy k,j}$ and $R_{yx k,j}$, for $k, j = 1, \ldots, N$ and $k \neq j$. In other words, in a general case, those covariances must be known.

Specially, in the case where the powers of all random processes are *equal* and other conditions hold as mentioned in Section 7.2.2 and 7.2.3, we can follow equations (7.3) and (7.4) in the case of time delay and frequency separation, such as in OFDM systems, or equations (7.5), (7.6) and (7.7) in the case of spatial separation like with multiple antennas in MIMO systems to calculate the covariances $R_{xx k,j}$, $R_{yy k,j}$, $R_{xy k,j}$ and $R_{yx k,j}$.

The values $\{\sigma_{g_j}^2\}_{j=1}^N$, $R_{xx k,j}$, $R_{yy k,j}$, $R_{xy k,j}$ and $R_{yx k,j}$ ($k, j = 1, \ldots, N; k \neq j$) are the input data of our proposed algorithm.

---

[3]Note that $\sigma_{g_j}^2$ is the variance of *complex* Gaussian random variables, rather than the variance per dimension (real or imaginary). Hence, there is no factor of 2 in the denominator of (7.14).

3 Create the $N \times N$-sized covariance matrix $\mathcal{K}$

$$\mathcal{K} = [\mu_{k,j}]_{N \times N} \qquad (7.15)$$

where

$$\mu_{k,j} = \begin{cases} \sigma_{g_j}^2 & \text{if } k \equiv j \\ (R_{xx\,k,j} + R_{yy\,k,j}) - i(R_{xy\,k,j} - R_{yx\,k,j}) & \text{if } k \neq j \end{cases} \qquad (7.16)$$

The covariance matrix of complex Gaussian random variables is considered here, as opposed to the covariance matrix of Rayleigh fading envelopes like in the conventional methods.

4 Perform the eigen decomposition

$$\mathcal{K} = \mathbf{V}\mathbf{G}\mathbf{V}^H$$

Denote $\mathbf{G} \triangleq diag(\lambda_1, \ldots, \lambda_N)$. Then, calculate a new diagonal matrix:

$$\mathbf{\Lambda} = diag(\hat{\lambda}_1, \ldots, \hat{\lambda}_N)$$

where

$$\hat{\lambda}_j = \begin{cases} \lambda_j & \text{if } \lambda_j \geq 0 \\ 0 & \text{if } \lambda_j < 0 \end{cases} \qquad j = 1, \ldots, N.$$

Thereby, we have a diagonal matrix $\mathbf{\Lambda}$ with all elements in the main diagonal being *real* and definitely *non-negative*.

5 Determine a new matrix $\bar{\mathbf{\Lambda}} = \sqrt{\mathbf{\Lambda}}$ and calculate the coloring matrix $\mathbf{L}$ by setting $\mathbf{L} = \mathbf{V}\bar{\mathbf{\Lambda}}$.

6 Generate a column vector $\mathbf{w}$ of N *independent* complex Gaussian random samples with zero means and *arbitrary, equal* variances σ_g^2

$$\mathbf{w} = (u_1, \ldots, u_N)^T$$

We can see that the modification 1 takes place in this step of our algorithm and proceeds in the next step.

7 Generate a column vector $\mathbf{z}$ of N *correlated* complex Gaussian random samples as follows:

$$\mathbf{z} = \frac{\mathbf{L}\mathbf{w}}{\sigma_g} \triangleq (z_1 \ldots, z_N)^T$$

As shown later in the next section, the elements $\{z_j\}_{j=1}^{N}$ are zero-mean, *correlated* complex Gaussian random variables with variances $\{\sigma_{g_j}^2\}_{j=1}^{N}$. The N moduli $\{r_j\}_{j=1}^{N}$ of the Gaussian samples in $\mathbf{z}$ are the *desired* Rayleigh fading envelopes.

7.2.4.5   Statistical Properties of the Resultant Envelopes

In this section, we check the covariance matrix and the variances (powers) of the resultant correlated complex Gaussian random samples as well as the variances (powers) of the resultant Rayleigh fading envelopes.

It is easy to see that $E(\mathbf{w}\mathbf{w}^H) = \sigma_g^2 \mathbf{I}_N$, and therefore

$$E(\mathbf{z}\mathbf{z}^H) \;=\; E\left(\frac{\mathbf{L}\mathbf{w}\mathbf{w}^H\mathbf{L}^H}{\sigma_g^2}\right) = E(\mathbf{L}\mathbf{L}^H) = \mathbf{K}$$

It means that the generated Rayleigh envelopes are corresponding to the forced positive semi-definite covariance matrix $\mathbf{K}$, which is, in turn, *equal* to the desired covariance matrix $\mathcal{K}$ in the case where $\mathcal{K}$ is *positive semi-definite*, or is a *well approximate* matrix of $\mathcal{K}$ otherwise. In other words, the desired covariance matrix $\mathcal{K}$ of complex Gaussian random variables (with Rayleigh fading envelopes) is achieved.

In addition, note that the variance of the jth Gaussian random variable in $\mathbf{z}$ is the jth element on the main diagonal of $\mathbf{K}$. Because $\mathbf{K}$ approximates to $\mathcal{K}$, the elements on the main diagonal of $\mathbf{K}$ are thus equal (or close) to $\sigma_{g_j}^2$s (see Eq. (7.15) and (7.16)). As a result, the resultant complex Gaussian random variables $\{z_j\}_{j=1}^{N}$ in $\mathbf{z}$ have zero means and variances (powers) $\{\sigma_{g_j}^2\}_{j=1}^{N}$.

It is known that the means and the variances of Rayleigh envelopes $\{r_j\}_{j=1}^{N}$ have the relation with the variances of the corresponding complex Gaussian random variables $\{z_j\}_{j=1}^{N}$ in $\mathbf{z}$ as given below (see (5.51), (5.52) in [Rappaport, 2002] and (2.1-131) in [Proakis, 2001])

$$E\{r_j\} \;=\; \sigma_{g_j}\frac{\sqrt{\pi}}{2} = 0.8862\sigma_{g_j}\,, \qquad (7.17)$$

$$Var\{r_j\} \;=\; \sigma_{g_j}^2\left(1 - \frac{\pi}{4}\right) = 0.2146\sigma_{g_j}^2 \qquad (7.18)$$

From (7.14), (7.17) and (7.18), it is clear that

$$E\{r_j\} \;=\; \sigma_{r_j}\sqrt{\frac{\pi}{4-\pi}}\,,$$

$$Var\{r_j\} \;=\; \sigma_{r_j}^2$$

Therefore, the *desired* variances (powers) $\{\sigma_{r_j}^2\}_{j=1}^{N}$ of Rayleigh envelopes are achieved.

7.2.5 Generation of Correlated Rayleigh Envelopes in a Real-Time Scenario

In Section 7.2.4.4, we have proposed the algorithm for generating N correlated Rayleigh fading envelopes in multipath, flat fading channels in a *single time instant*. We can repeat steps 6 and 7 of this algorithm to generate Rayleigh envelopes in the *continuous time interval*. It is noted that, the discrete-time samples of each Rayleigh fading process generated by this algorithm in *different* time instants are *independent* of each other.

It has been known that the discrete-time samples of each *realistic* Rayleigh fading process may have *autocorrelation* properties, which are the functions of the Doppler frequency corresponding to the motion of receivers as well as other factors such as the sampling frequency of transmitted signals. It is because the band-limited communication channels not only limit the bandwidth of transmitted signals, but also limit the bandwidth of fading. This filtering effect limits the rate of changes of fading in time domain, and consequently, results in the autocorrelation properties of fading. Therefore, the algorithm generating Rayleigh fading envelopes in *realistic* conditions must consider the autocorrelation properties of Rayleigh fading envelopes.

To simulate a realistic multipath fading channel, Doppler filters are normally used [Rappaport, 2002]. The analysis of Doppler spectrum spread was first derived by Gans [Gans, 1972], based on Clarke's model [Clarke, 1968]. Motivated by these works, Smith [Smith, 1975] developed a computer-assisted model generating an *individual* Rayleigh fading envelope in flat fading channels corresponding to a given *normalized autocorrelation* function. This model was then modified by Young [Young and Beaulieu, 1996], [Young and Beaulieu, 2000] to provide more accurate channel realization.

It should be emphasized that, in [Young and Beaulieu, 1996], [Young and Beaulieu, 2000], the models are aimed at generating an *individual* Rayleigh envelope corresponding to a certain normalized *autocorrelation* function of itself, rather than generating different Rayleigh envelopes corresponding to a desired covariance matrix (*autocorrelation* and *cross-correlation* properties between those envelopes).

Therefore, the model for generating N correlated Rayleigh fading envelopes in realistic fading channels (each individual envelope is corresponding to a desired normalized autocorrelation property) can be created by associating the model proposed in [Young and Beaulieu, 2000] with the algorithm mentioned

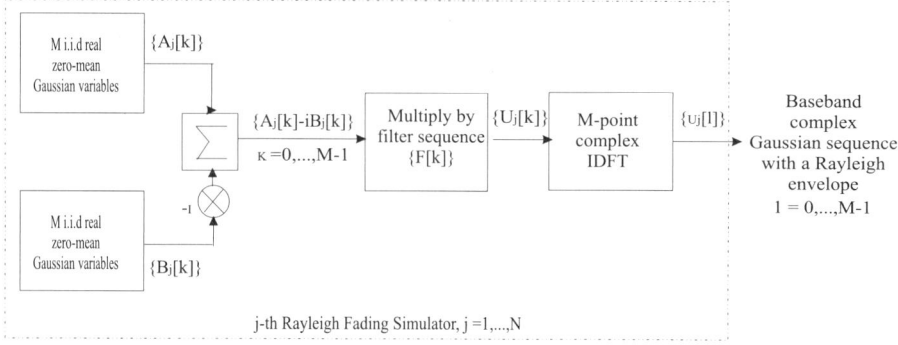

Figure 7.2. Model of a Rayleigh generator for an individual Rayleigh envelope corresponding to a desired *normalized* autocorrelation function.

in Section 7.2.4.4 in such a way that, the resultant Rayleigh fading envelopes are corresponding to the desired covariance matrix.

This combination must overcome the main shortcoming of the method proposed in [Sorooshyari and Daut, 2003] as analyzed in Section 7.2.1. In other words, the modification 2 mentioned in Section 7.2.4.4 must be carried out. This is an easy task in our algorithm. The key for the success of this task is the modification in steps 6 and 7 of our algorithm (see Section 7.2.4.4), where the variances of N complex Gaussian random variables are *not fixed* as in [Sorooshyari and Daut, 2003], but can be *arbitrary* in our algorithm. Again, besides being more generalized, our modification in these steps allows the *accurate* combination of the method proposed in [Young and Beaulieu, 2000] and our algorithm, i.e. guaranteeing that the generated Rayleigh envelopes are exactly corresponding to the desired covariance matrix.

The model of a Rayleigh fading generator for generating an *individual* baseband Rayleigh fading envelope proposed in [Young and Beaulieu, 1996], [Young and Beaulieu, 2000] is shown in Fig. 7.2. This model generates a Rayleigh fading envelope using Inverse Discrete Fourier Transform (IDFT), based on *independent* zero-mean Gaussian random variables weighted by appropriate Doppler filter coefficients. The sequence $\{u_j[l]\}_{l=0}^{M-1}$ of the complex Gaussian random samples at the output of the jth Rayleigh generator (Fig. 7.2) can be expressed as

$$u_j[l] = \frac{1}{M} \sum_{k=0}^{M-1} U_j[k] e^{i\frac{2\pi k l}{M}}$$

where

- M denotes the number of points with which the IDFT is carried out;

- l is the discrete-time sample index ($l = 0, \ldots, M - 1$);

- $U_j[k] = F[k]A_j[k] - iF[k]B_j[k]$;

- $\{F[k]\}$ are the Doppler filter coefficients.

For brevity, we omit the subscript j in the expressions, except when this subscript is necessary to emphasize. If we denote $u[l]=u_R[l] + iu_I[l]$, then it has been proved that, the *autocorrelation* property between the real parts $u_R[l]$ and $u_R[m]$ as well as that between the imaginary parts $u_I[l]$ and $u_I[m]$ at different discrete-time instants l and m are as given below (see Eq. (7) in [Young and Beaulieu, 2000]):

$$
\begin{aligned}
r_{RR}[l,m] &= r_{II}[l,m] = r_{RR}[d] = r_{II}[d] \\
&= E\{u_R[l]u_R[m]\} = \frac{\sigma_{orig}^2}{M} Re\{g[d]\}
\end{aligned} \tag{7.19}
$$

where $d \triangleq l - m$ is the sample lag, σ_{orig}^2 is the variance of the *real*, independent zero-mean Gaussian random sequences $\{A[k]\}$ and $\{B[k]\}$ at the inputs of Doppler filters, and the sequence $\{g[d]\}$ is the IDFT of $\{F[k]^2\}$, i.e.

$$
g[d] = \frac{1}{M} \sum_{k=0}^{M-1} F[k]^2 e^{i\frac{2\pi kd}{M}} \tag{7.20}
$$

Similarly, the correlation property between the real part $u_R[l]$ and the imaginary part $u_I[m]$ is calculated as (see Eq. (8) in [Young and Beaulieu, 2000])

$$
r_{RI}[d] = E\{u_R[l]u_I[m]\} = \frac{\sigma_{orig}^2}{M} Im\{g[d]\} \tag{7.21}
$$

The mean value of the output sequence $\{u[l]\}$ has been proved to be zero (see Appendix A in [Young and Beaulieu, 2000]).

If $d=0$ and $\{F[k]\}$ are real, from (7.19), (7.20) and (7.21), we have

$$
\begin{aligned}
r_{RR}[0] &= r_{II}[0] = E\{u_R[l]u_R[l]\} = \frac{\sigma_{orig}^2}{M^2} \sum_{k=0}^{M-1} F[k]^2 , \\
r_{RI}[0] &= E\{u_R[l]u_I[l]\} = 0
\end{aligned} \tag{7.22}
$$

Therefore, by definition, the variance of the sequence $\{u[l]\}$ at the output of the Rayleigh generator is

$$\sigma_g^2 \triangleq E\{u[l]u[l]^*\} = 2E\{u_R[l]u_R[l]\} = \frac{2\sigma_{orig}^2}{M^2} \sum_{k=0}^{M-1} F[k]^2 \qquad (7.23)$$

where * denotes the complex conjugate operation.

Let r_{nor} be

$$r_{nor} = \frac{r_{RR}[d]}{\sigma_g^2} = \frac{r_{II}[d]}{\sigma_g^2} , \qquad (7.24)$$

i.e. let r_{nor} be the autocorrelation function in (7.19) *normalized* by the variance σ_g^2 in (7.23). r_{nor} is called the *normalized autocorrelation* function.

To achieve a desired *normalized autocorrelation* function $r_{nor} = J_0(2\pi f_m d)$, where f_m is the maximum Doppler frequency F_m normalized by the sampling frequency F_s of the transmitted signals (i.e. $f_m = \frac{F_m}{F_s}$), the Doppler filter coefficients $\{F[k]\}$ are determined in Young's model [Young and Beaulieu, 1996], [Young and Beaulieu, 2000] as follows (see Eq. (21) in [Young and Beaulieu, 2000]):

$$F[k] = \begin{cases} 0 & k = 0 \\ \sqrt{\dfrac{1}{2\sqrt{1-\left(\frac{k}{Mf_m}\right)^2}}} & k = 1, \ldots, k_m - 1 \\ \sqrt{\dfrac{k_m}{2}\left[\frac{\pi}{2} - \arctan\left(\frac{k_m-1}{\sqrt{2k_m-1}}\right)\right]} & k = k_m \\ 0 & k = k_m + 1, \ldots, M - k_m - 1 \\ \sqrt{\dfrac{k_m}{2}\left[\frac{\pi}{2} - \arctan\left(\frac{k_m-1}{\sqrt{2k_m-1}}\right)\right]} & k = M - k_m \\ \sqrt{\dfrac{1}{2\sqrt{1-\left(\frac{M-k}{Mf_m}\right)^2}}} & k = M - k_m + 1, \ldots, M - 2, M - 1 \end{cases} \qquad (7.25)$$

In (7.25), $k_m \triangleq \lfloor f_m M \rfloor$, where $\lfloor . \rfloor$ indicates the biggest rounded integer being less or equal to the argument.

It has been proved in [Young and Beaulieu, 2000] that the real filter coefficients in (7.25) will produce a complex Gaussian sequence with the *normalized autocorrelation* function $J_0(2\pi f_m d)$, where d is the sample lag, and with the *expected independence* between the real and imaginary parts of Gaussian samples, i.e., the correlation property in (7.21) is zero. The zero-correlation property between the real and imaginary parts is necessary in order that the resultant envelopes are Rayleigh distributed.

Let us consider the variance σ_g^2 of the resultant complex Gaussian sequence at the output of Fig. 7.2. We consider an example where $M = 4096$, $f_m = 0.05$

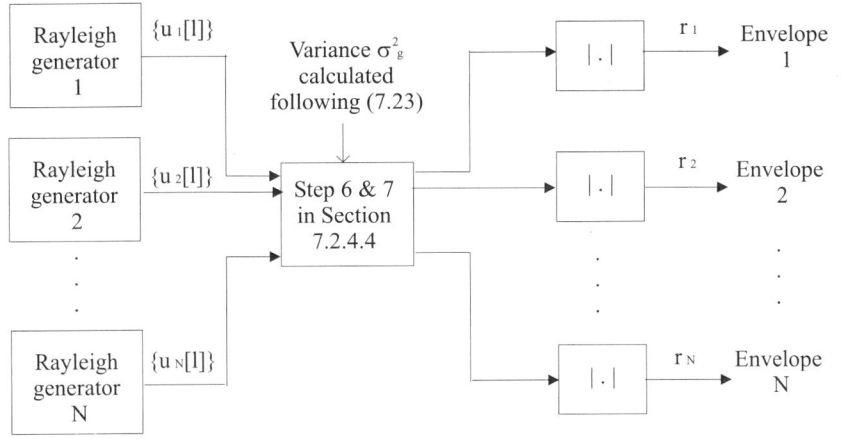

Figure 7.3. Model for generating N Rayleigh fading envelopes corresponding to a desired *normalized* autocorrelation function in a real-time scenario.

and $\sigma^2_{orig} = 1/2$ (σ^2_{orig} is the variance per dimension). From (7.23) and (7.25), we have $\sigma^2_g = 1.8965.10^{-5}$. Clearly, passing complex Gaussian random variables with unit variances through Doppler filters reduces significantly the variances of those variables. In general, the variances of the complex Gaussian random variables at the output of the Rayleigh simulator presented in Fig. 7.2 can be *arbitrary*, depending on M, σ^2_{orig} and $\{F[k]\}$, i.e. depending on the variances of the Gaussian random variables at the inputs of the Doppler filters as well as the characteristics of those filters (see Eq. (7.23) for more details).

We now return to the main shortcoming of the method proposed in [Sorooshyari and Daut, 2003], which was mentioned earlier in Section 7.2.1. In Section VI of the paper [Sorooshyari and Daut, 2003], the authors generated Rayleigh envelopes corresponding to a desired covariance matrix in a *real-time scenario*, where Doppler frequency shifts were considered, by combining their proposed method with the method in [Young and Beaulieu, 2000]. Specifically, the authors took the outputs of the method in [Young and Beaulieu, 2000] and *simply* input them into step 6 in their algorithm.

However, the step 6 in the method in [Sorooshyari and Daut, 2003] was proposed for generating complex Gaussian random variables with a *fixed* (unit) variance. Meanwhile, as presented earlier, the variances of the complex Gaussian random variables at the output of the Rayleigh simulator may have *arbitrary* values, depending on the variances of the Gaussian random variables at the inputs of the Doppler filters as well as the characteristics of those filters. Consequently, if the outputs of the method in [Young and Beaulieu, 2000] are simply input into the step 6 as mentioned in the algorithm in [Sorooshyari and Daut, 2003], the covariance matrix of the resultant correlated Gaussian random variables is *not* equal to the desired covariance matrix due to the variance-changing effect of Doppler filters being *not* considered. In other words, the method proposed in [Sorooshyari and Daut, 2003] *fails* to generate Rayleigh fading envelopes corresponding to a desired covariance matrix in a real-time scenario where Doppler frequency shifts are taken into account.

Our model for generating N correlated Rayleigh fading envelopes corresponding to a desired covariance matrix in a real-time scenario where Doppler frequency shifts are considered is presented in Fig. 7.3. In this model, N Rayleigh generators, each of which is presented in Fig. 7.2, are simultaneously used. To generate N *correlated* Rayleigh envelopes corresponding to a desired covariance matrix at *an observed discrete-time instant* l ($l = 0, \ldots, M - 1$), similarly to the method in [Sorooshyari and Daut, 2003], we take the output $u_j[l]$ of the jth Rayleigh simulator, for $j = 1, \ldots, N$, and input it as the element u_j into step 6 of our algorithm proposed in Section 7.2.4.4. However, as opposed to the method in [Sorooshyari and Daut, 2003], the variance σ_g^2 of complex Gaussian samples u_j in step 6 of our method is calculated according to (7.23). This value is used as the input parameter for steps 6 and 7 of our algorithm (see Fig. 7.3). Thereby, the variance-changing effect caused by Doppler filters is taken into consideration in our algorithm.

The algorithm for generating N correlated Rayleigh envelopes (when Doppler frequency shifts are considered) *at a discrete time instant* l, for $l = 0, \ldots, M - 1$, can be summarized as:

1 Perform the steps 1 to 5 mentioned in Section 7.2.4.4.

2 From the *desired autocorrelation* properties (7.19) and (7.24) of each of the complex Gaussian random sequences (with Rayleigh fading envelopes), determine the values M and σ_{orig}^2. These values can be arbitrarily selected, provided that they bring about the desired autocorrelation properties. The value of M is also the number of points with which the IDFT is carried out.

3 For each Rayleigh generator presented in Fig. 7.2, generate M i.i.d., *real*, zero-mean Gaussian random samples $\{A[k]\}$ with the variance σ_{orig}^2 and, independently, generate M i.i.d., *real*, zero-mean Gaussian samples $\{B[k]\}$ with the distribution $(0, \sigma_{orig}^2)$. From $\{A[k]\}$ and $\{B[k]\}$, generate M i.i.d complex Gaussian random variables $\{A[k] - iB[k]\}$. N Rayleigh generators are simultaneously used to generate N Rayleigh envelopes as presented in Fig. 7.3.

4 Multiply complex Gaussian samples $\{A[k] - iB[k]\}$, for $k = 0, \ldots, M-1$, with the corresponding filter coefficient $F[k]$ given in (7.25).

5 Perform M-point IDFT of the resultant samples.

6 Calculate the variance σ_g^2 of the output $\{u[l]\}$ following (7.23). It is noted that σ_g^2 is the same for N Rayleigh generators. We also emphasize that, by this calculation, the modification 2 mentioned in Section 7.2.4.4 has been performed in this step.

7 Create a column vector $\mathbf{w} = (u_1, \ldots, u_N)^T$ of N i.i.d. complex Gaussian random samples with the distribution $(0, \sigma_g^2)$ where the element u_j, for $j = 1, \ldots, N$, is the output $u_j[l]$ of the jth Rayleigh generator and σ_g^2 has been calculated in step 6.

8 Continue the step 7 mentioned in Section 7.2.4.4. The N envelopes of elements in the column vector $\mathbf{z}$ are the desired Rayleigh envelopes *at the considered time instant l.*

Steps 7 and 8 are repeated for different time instants l ($l = 0, \ldots, M-1$), and therefore, the algorithm can be used for a real-time scenario.

7.2.6   Simulation Results

In this section, first, we simulate $N=3$ *frequency-correlated* Rayleigh fading envelopes corresponding to the complex Gaussian random variables with equal powers $\sigma_{g_j}^2 = 1$ ($j = 1, \ldots, 3$) in the flat fading channels. Parameters considered here include $M = 2^{14}$ (the number of IDFT points), $\sigma_{orig}^2 = 1/2$ (variances per dimension in Young's model), $F_s = 8$ KHz, $F_m = 50$ Hz (corresponding to a carrier frequency of 900 MHz and a mobile speed of $v = 60$ km/hr). Frequency separation between two adjacent carrier frequencies considered here is $\Delta f = 200$ kHz (e.g in GSM 900) and we assume that $f_1 > f_2 > f_3$. Also, we consider the rms delay spread $\sigma_\tau = 1\mu s$ and time delays between three envelopes are $\tau_{1,2} = 1ms$, $\tau_{2,3} = 3ms$, $\tau_{1,3} = 4ms$.

From (7.3), (7.4), (7.15) and (7.16), we have the *desired* covariance matrix $\mathcal{K}$ as given below:

$$\mathcal{K} = \begin{bmatrix} 1 & 0.3782 + 0.4753i & 0.0878 + 0.2207i \\ 0.3782 - 0.4753i & 1 & 0.3063 + 0.3849i \\ 0.0878 - 0.2207i & 0.3063 - 0.3849i & 1 \end{bmatrix} \quad (7.26)$$

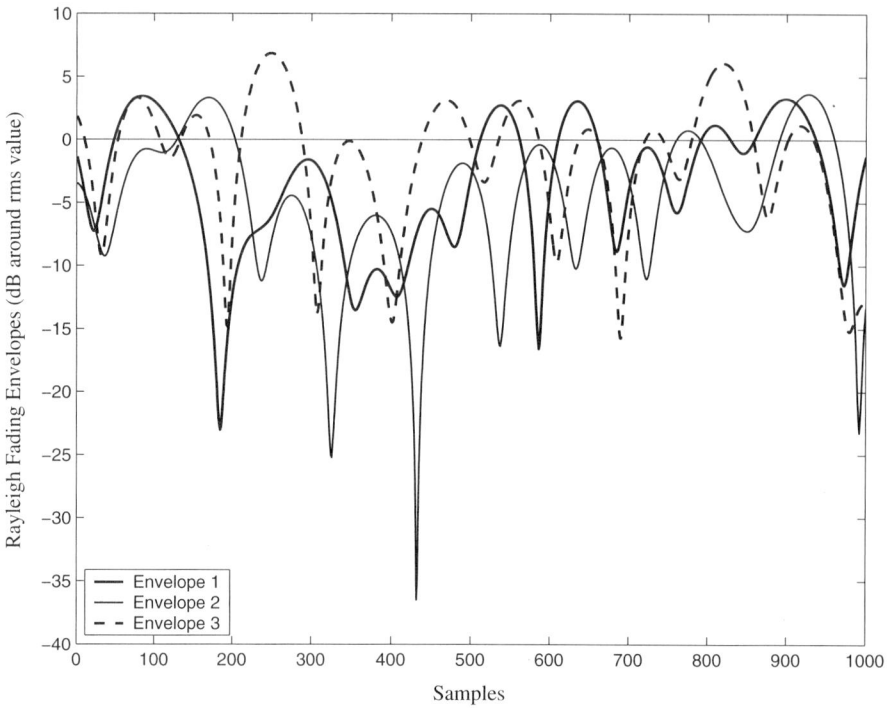

Figure 7.4. Example of three equal power, *spectrally correlated* Rayleigh fading envelopes with GSM specifications.

It is easy to check that $\mathcal{K}$ in (7.26) is positive definite. Using the proposed algorithm in Section 7.2.5, we have the simulation result presented in Fig. 7.4.

Next, we simulate N=3 *spatially-correlated* Rayleigh fading envelopes. We consider an antenna array comprising three Tx antennas, which are equally separated by a distance D. Assume that $\frac{D}{\lambda} = 1$, i.e., $D = 33.3$ cm for GSM 900. Additionally, we assume that $\Delta = \pi/18$ rad (or $\Delta = 10°$) and $\Phi = 0$ rad. The parameters M, $\sigma_{g_j}^2$, σ_{orig}^2, F_s and F_m are the same as in the previous case.

From (7.5), (7.6), (7.7), (7.15) and (7.16), we have the following *desired* covariance matrix:

$$\mathcal{K} = \begin{bmatrix} 1 & 0.8123 & 0.3730 \\ 0.8123 & 1 & 0.8123 \\ 0.3730 & 0.8123 & 1 \end{bmatrix} \qquad (7.27)$$

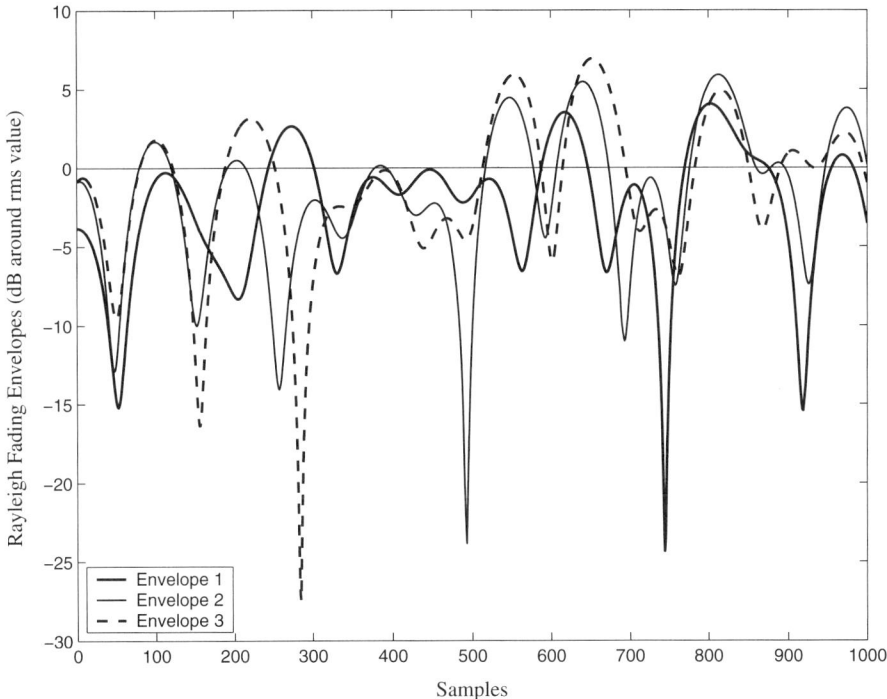

Figure 7.5. Example of three equal power, *spatially correlated* Rayleigh fading envelopes with GSM specifications.

Since $\Phi = 0$ rad, the covariances $R_{xy_{k,j}}$ and $R_{yx_{k,j}}$ between the real and imaginary components of any pair of the complex Gaussian random processes (with Rayleigh fading envelopes) are zero, and consequently, $\mathcal{K}$ is a *real* matrix. Readers may refer to (7.6) and (7.7) for more details. It is easy to realize that $\mathcal{K}$ in (7.27) is positive definite. The simulation result is presented in Fig. 7.5.

In Fig. 7.6, we simulate $N=3$ *frequency-correlated* Rayleigh envelopes based on IEEE 802.11a (OFDM) specifications [Standards Association, 1999]. In particular, the parameters considered here include $M = 2^{20}$, $\sigma_{g_j}^2 = 1$ $(j = 1, \ldots, 3)$, $\sigma_{orig}^2 = 1/2$, $F_s = 20$ MHz, $F_m = 555.56$ Hz (corresponding to a carrier frequency of 5 GHz and a mobile speed of $v = 120$ km/hr), $\Delta f = 312.5$ kHz, $\sigma_\tau = 0.1\mu s$, $\tau_{1,2} = \tau_{2,3} = 1ms$, and $\tau_{1,3} = 2ms$. In Fig. 7.7, we simulate the case where the covariance matrix is not positive semi-definite as mentioned earlier in Example 7.2.1 of Section 7.2.4.1. From Fig. 7.7, we can realize that the three Rayleigh envelopes are highly correlated as we expect (see Eq. (7.10)).

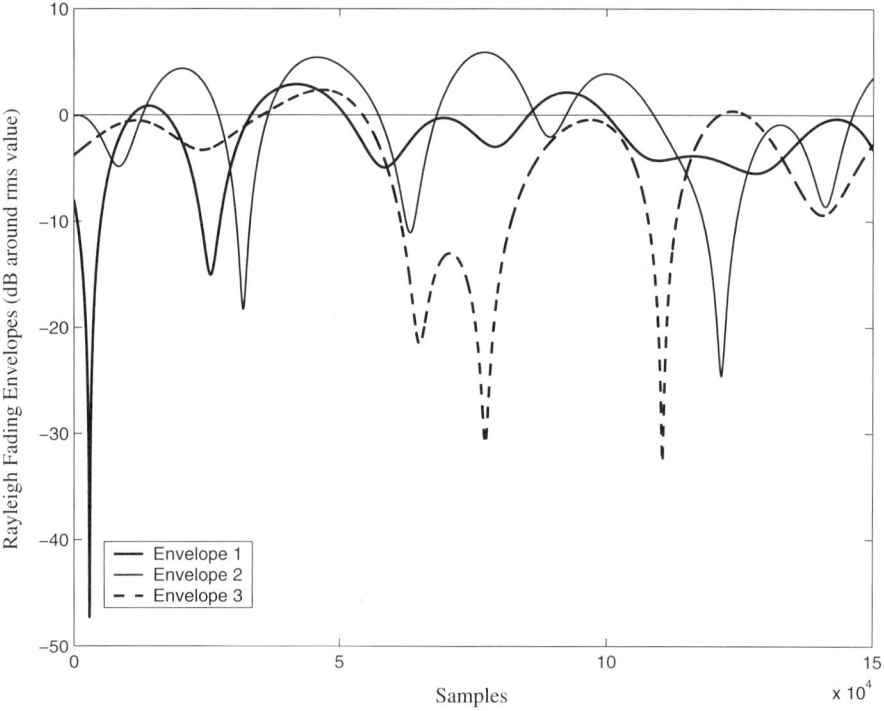

Figure 7.6. Example of three equal power, *spectrally correlated* Rayleigh fading envelopes with IEEE 802.11a (OFDM) specifications.

In Fig. 7.8, we plot the histograms of the resultant Rayleigh fading envelopes produced by our algorithm in the four aforementioned examples. Without loss of generality, we plot the histograms for one of three Rayleigh fading envelopes, such as the first Rayleigh fading envelope. To compare the accuracy of our algorithm, we also plot the *theoretical* probability density function (PDF) of a typical Rayleigh fading envelope by solid curves. In this figure, the parameter $\sigma_{g_j}^2$ of the PDF is the variance of the *complex* Gaussian random process corresponding to the considered typical Rayleigh fading envelope. It can be observed from Fig. 7.8 that the resultant envelopes produced by our algorithm in the four examples follow accurately the theoretical PDF of the typical Rayleigh fading envelope.

Finally, in Fig. 7.9, we compare the computational efforts between our algorithm and the one mentioned in [Sorooshyari and Daut, 2003], by comparing the average computational time required for both algorithms to simulate $N =$

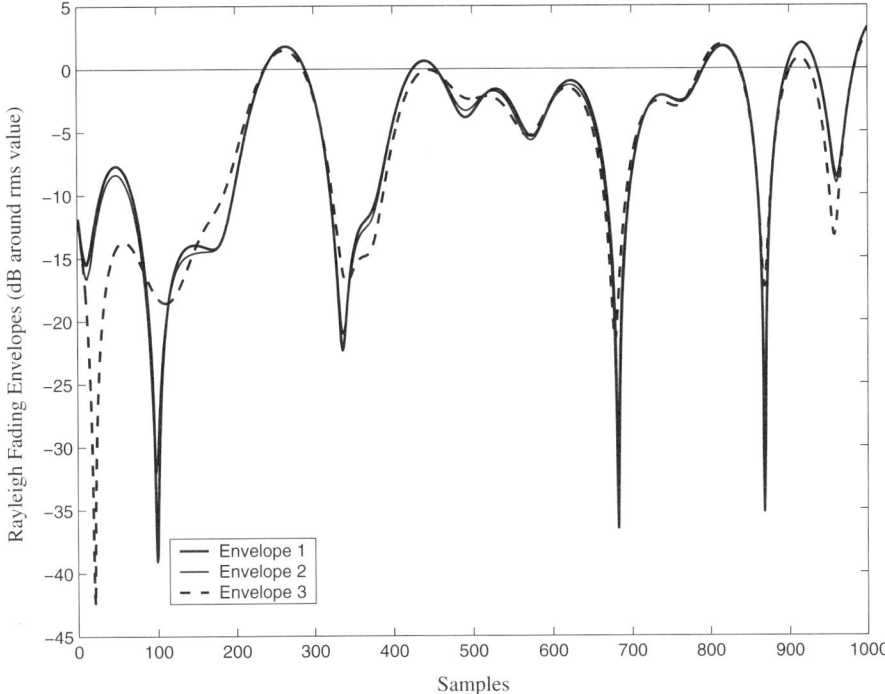

Figure 7.7. Example of three equal power, *spatially correlated* Rayleigh fading envelopes with a not positive semi-definite covariance matrix.

2, 4, 8, 16, 32, 64 or 128 Rayleigh envelopes in a real-time scenario over 10000 trials. It can be realized from Fig. 7.9 that, for $N = 64$ and $N = 128$, our algorithm is slightly more complex, while it is almost as computationally efficient as the method in [Sorooshyari and Daut, 2003] for a smaller N.

7.3 Performance of Diversity Antenna Selection Techniques in Correlated, Flat Rayleigh Fading Channels Using DSTBCs

7.3.1 AST/DSTBC Schemes in Correlated, Flat Rayleigh Fading Channels

In Section 6.3, we have proposed two ASTs, which are referred to as the *general* $(M, N; K)$ AST/DSTBC and the *restricted* $(M, N; K)$ AST/DSTBC schemes, for channels utilizing DSTBCs. It has been shown there that the proposed AST/DSTBC schemes improve significantly the bit error performance of channels using DSTBCs. This conclusion is derived from the assumption

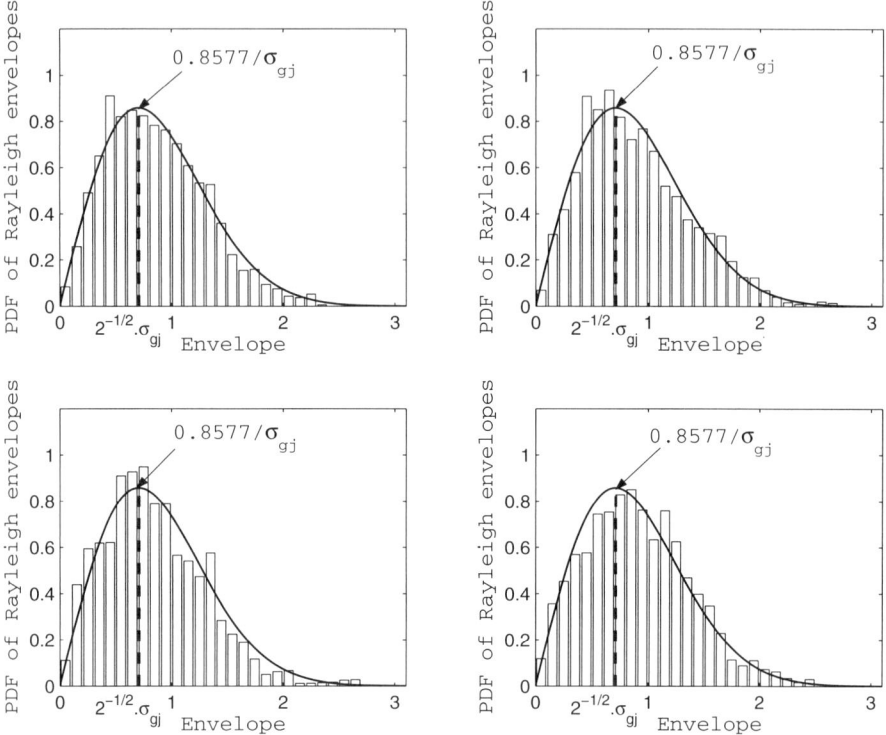

Figure 7.8. Histograms of Rayleigh fading envelopes produced by the proposed algorithm in the four examples along with a Rayleigh PDF where $\sigma_{g_j}^2 = 1$.

that the channels are independent Rayleigh flat fading ones. This assumption is valid, for instance, when Tx antennas as well as Rx antennas are sufficiently spatially separated from one another. The distance between the two adjacent antennas are normally selected to be multiple of half of the wavelength.

In fact, the spatial separation between the transmission antennas may not be sufficient resulting in correlation between those antennas. In such a scenario, we may wonder whether the superiority of the proposed AST/DSTBC schemes still holds. Therefore, in this section, we consider the AST/DSTBC schemes proposed for systems using DSTBCs in *spatially* correlated, flat Rayleigh fading channels.

Specifically, we first examine wireless systems comprising 1 Rx antenna and using DSTBCs based on the Alamouti code [Alamouti, 1998] (or Alamouti DSTBC, for short) in association with the *general* (3,2;1) AST/DSTBC and the *restricted* (3,2;1) AST/DSTBC schemes. We then examine wireless systems us-

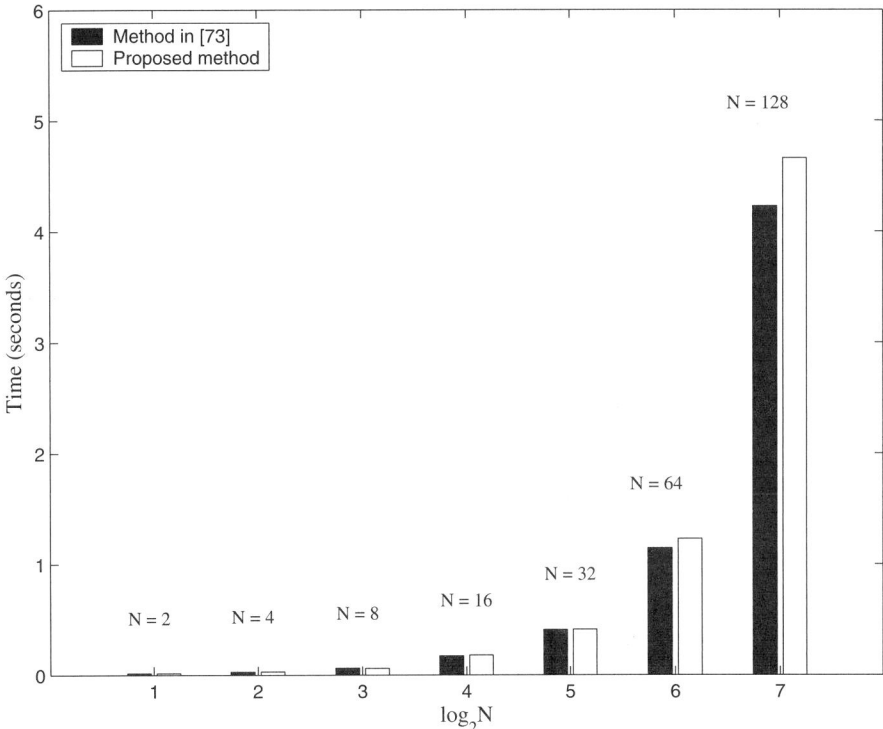

Figure 7.9. Computational effort comparison between the method in [Sorooshyari and Daut, 2003] and the proposed algorithm.

ing the Alamouti DSTBC associated with the *general* (4,2;1) AST/DSTBC and the *restricted* (4,2;1) AST/DSTBC schemes to confirm further the superiority of our proposed AST/DSTBC schemes in the correlated Rayleigh fading scenario. Systems using the Alamouti DSTBC with no ASTs are also considered here. They are used as references to evaluate the superiority of our proposed AST/DSTBC schemes.

We have three different values of the total number M of Tx antennas for all the cases considered here. Particularly, the system may comprise:

- $M=2$ Tx antennas corresponding to the case of the Alamouti DSTBC without AST/DSTBC schemes.

- $M=3$ Tx antennas in the case of the Alamouti DSTBC in association with the *general* (3,2;1) AST/DSTBC scheme or with the *restricted* (3,2;1) AST/DSTBC scheme.

- M=4 Tx antennas in the case of the Alamouti DSTBC in association with the *general* (4,2;1) AST/DSTBC scheme or with the *restricted* (4,2;1) AST/DSTBC scheme.

In the following expressions, $M \times M$-sized matrices (M=2, 3 or 4) present the covariance matrices of the M transmission coefficients (between M Tx antennas and the Rx antenna). The transmission coefficients are assumed to be zero-mean, complex, correlated Gaussian random variables with Rayleigh envelopes. Further, we assume that those Gaussian random variables have the same powers $\sigma_{g_j}^2 = 1$, for $j = 1, \ldots, M$.

We examine the two following cases. One is corresponding to a low correlation level, and, the other is corresponding to a high correlation level. Particularly, in the case of the low correlation level, we assume that $\frac{D}{\lambda} = 2$, i.e., $D = 66.6$ cm for GSM 900. Additionally, we assume that $\Delta = \pi/2$ rad and $\Phi = 0$ rad. Readers may refer to Section 7.2.3 and Fig. 7.1 for the definition and the illustration of these notations.

From (7.5), (7.6), (7.7), (7.15) and (7.16), we have the following desired covariance matrices:

- For $M = 2$:

$$\mathcal{K}_1 = \begin{bmatrix} 1 & 0.1575 \\ 0.1575 & 1 \end{bmatrix} \tag{7.28}$$

- For $M = 3$:

$$\mathcal{K}_2 = \begin{bmatrix} 1 & 0.1575 & 0.1120 \\ 0.1575 & 1 & 0.1575 \\ 0.1120 & 0.1575 & 1 \end{bmatrix} \tag{7.29}$$

- For $M = 4$:

$$\mathcal{K}_3 = \begin{bmatrix} 1 & 0.1575 & 0.1120 & 0.0916 \\ 0.1575 & 1 & 0.1575 & 0.1120 \\ 0.1120 & 0.1575 & 1 & 0.1575 \\ 0.0916 & 0.1120 & 0.1575 & 1 \end{bmatrix} \tag{7.30}$$

In the case of the high correlation level, we assume that $\frac{D}{\lambda} = 1$, i.e., $D = 33.3$ cm for GSM 900. Further, we assume that $\Delta = \pi/18$ rad (or $\Delta = 10°$) and $\Phi = 0$ rad. From (7.5), (7.6), (7.7), (7.15) and (7.16), we have the following desired covariance matrices:

- For $M = 2$:

$$\mathcal{K}_4 = \begin{bmatrix} 1 & 0.8123 \\ 0.8123 & 1 \end{bmatrix} \tag{7.31}$$

- For $M = 3$:

$$\mathcal{K}_5 = \begin{bmatrix} 1 & 0.8123 & 0.3730 \\ 0.8123 & 1 & 0.8123 \\ 0.3730 & 0.8123 & 1 \end{bmatrix} \tag{7.32}$$

- For $M = 4$:

$$\mathcal{K}_6 = \begin{bmatrix} 1 & 0.8123 & 0.3730 & -0.0432 \\ 0.8123 & 1 & 0.8123 & 0.3730 \\ 0.3730 & 0.8123 & 1 & 0.8123 \\ -0.0432 & 0.3730 & 0.8123 & 1 \end{bmatrix} \tag{7.33}$$

It is easy to realize that all of the above covariance matrices are positive definite. For ease of exposition, let $\mathcal{K}_A$ and $\mathcal{K}_B$ be the sets of the covariance matrices $\{\mathcal{K}\}_{j=1}^{3}$ and $\{\mathcal{K}\}_{j=4}^{6}$, respectively. These covariance matrices are utilized to simulate correlated flat Rayleigh fading channels which are differentially space-time coded and equipped with our proposed AST/DSTBC schemes.

7.3.2   Simulation Results

In this section, a wireless system utilizing the Alamouti DSTBC and comprising M=2, 3 or 4 Tx antennas and 1 Rx antenna is considered. Since the Alamouti DSTBC is used, the *required* number of Tx antennas is N=2. The redundant Tx antennas are used to provide spatial diversity.

The correlated *Rayleigh* transmission coefficients are generated based on the algorithm mentioned in Section 7.2.4.4 and corresponding to the sets $\mathcal{K}_A$ and $\mathcal{K}_B$ of the covariance matrices mentioned in (7.28) – (7.30), and (7.31) – (7.33), respectively. Noise is assumed to be i.i.d. complex Gaussian random variables with the distribution $\mathcal{CN}(0, \sigma_N^2)$. In the simulation, SNR means the channel signal-to-noise ratio, i.e., the ratio between the total power of the received signals during each symbol time slot (STS) and the noise power. The transmitted symbols in the Alamouti DSTBC are modulated by a unit-power QPSK signal constellation. As a result, the average power of the received signals during each STS is one. Therefore, we assign the noise variance to $\sigma_N^2 = 1/SNR$ in order for SNR to be the channel signal-to-noise ratio.

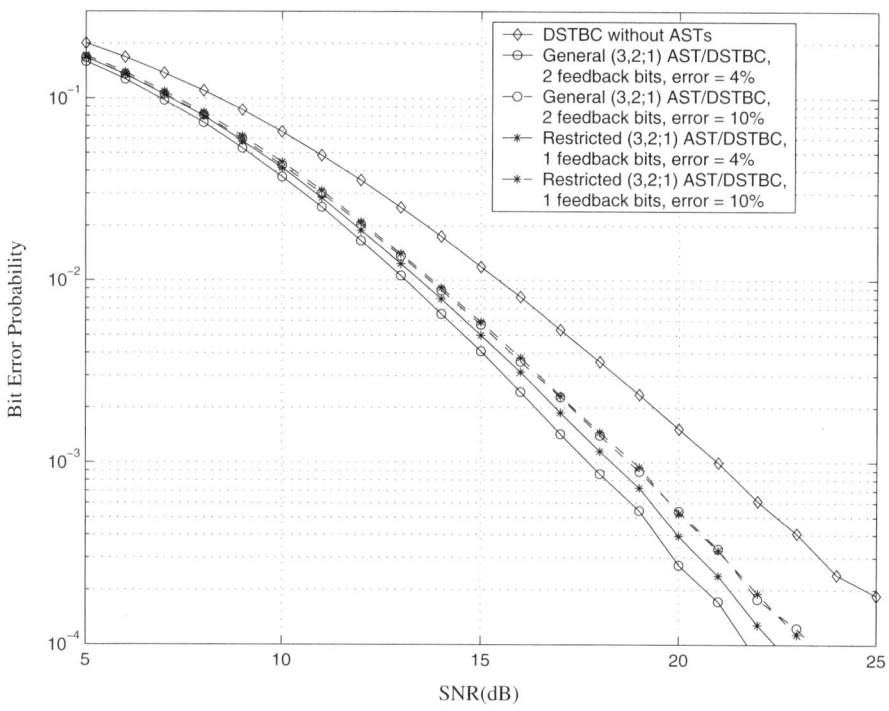

Figure 7.10. Proposed (3,2;1) AST/DSTBC schemes in correlated Rayleigh fading channels corresponding to the set $\mathcal{K}_A$ of covariance matrices.

Fig. 7.10 and Fig. 7.11 present the bit error rates (BER) of the Alamouti DSTBC associated with the *general* (3,2;1) AST/DSTBC and with the *restricted* (3,2;1) AST/DSTBC in correlated Rayleigh fading channels corresponding to the sets $\mathcal{K}_A$ and $\mathcal{K}_B$ of the covariance matrices, respectively. The solid lines are corresponding to the feedback error rate of 4%, while the dashed lines are corresponding to the feedback error rate of 10%.

Fig. 7.10 shows that, in the correlated flat Rayleigh fading channels corresponding to the set $\mathcal{K}_A$, the Alamouti DSTBC associated with our proposed AST/DSTBC shemes still overwhelms that without AST/DSTBC shemes. In particular, to achieve the same $BER = 10^{-3}$, the SNR gains achieved by the *general* (3,2;1) AST/DSTBC and the *restricted* (3,2;1) AST/DSTBC are 3.25 dB and 2.7 dB, respectively, with the feedback error rate of 4% and 2.25 dB and 2.20 dB, respectively, with the feedback error rate of 10%, compared to the case of the Alamouti DSTBC without our AST/DSTBC schemes.

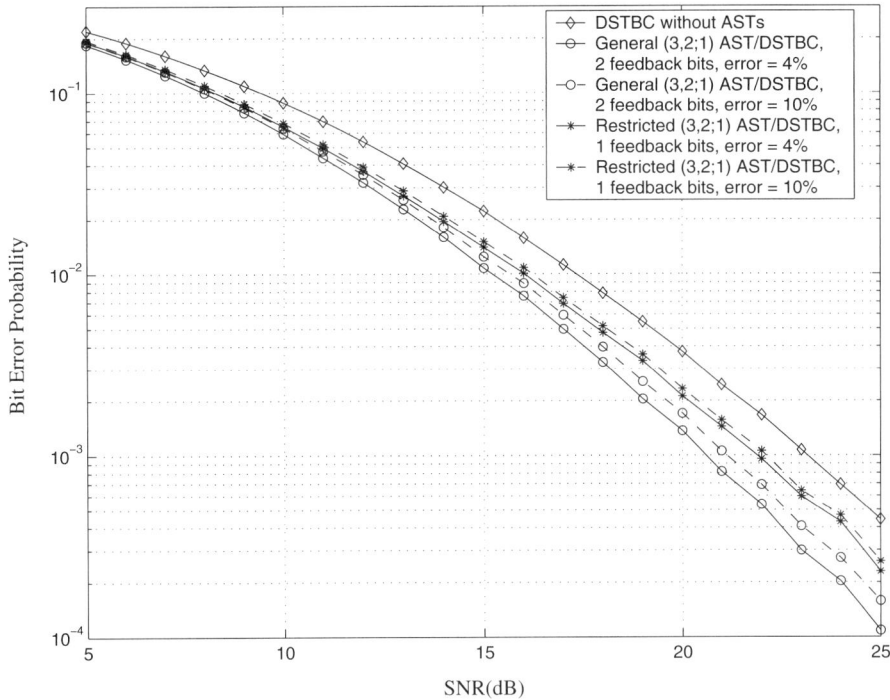

Figure 7.11. Proposed (3,2;1) AST/DSTBC schemes in correlated Rayleigh fading channels corresponding to the set $\mathcal{K}_B$ of covariance matrices.

Similarly, Fig. 7.11 shows that, in the more correlated fading channels corresponding to the set $\mathcal{K}_B$, the superiority of our proposed ASTs still holds. Specifically, to achieve the same $BER=10^{-3}$, the SNR gains achieved by the *general* (3,2;1) AST/DSTBC and the *restricted* (3,2;1) AST/DSTBC are 2.6 dB and 1.25 dB, respectively, with the feedback error rate of 4% and 2 dB and 1 dB, respectively, with the feedback error rate of 10%, compared to the case of the Alamouti DSTBC without our AST/DSTBC schemes.

Fig. 7.12 and Fig. 7.13 present the BER of the Alamouti DSTBC associated with the *general* (4,2;1) AST/DSTBC and with the *restricted* (4,2;1) AST/DSTBC in correlated Rayleigh fading channels corresponding to the sets $\mathcal{K}_A$ and $\mathcal{K}_B$, respectively.

We can realize from Fig. 7.12 that, in the correlated flat Rayleigh fading channels corresponding to the set $\mathcal{K}_A$, to achieve the same $BER = 10^{-3}$, the SNR gains achieved by the *general* (4,2;1) AST/DSTBC is 3.3 dB with the feedback error rate of 4% and 1.25 dB with the feedback error rate of 10%,

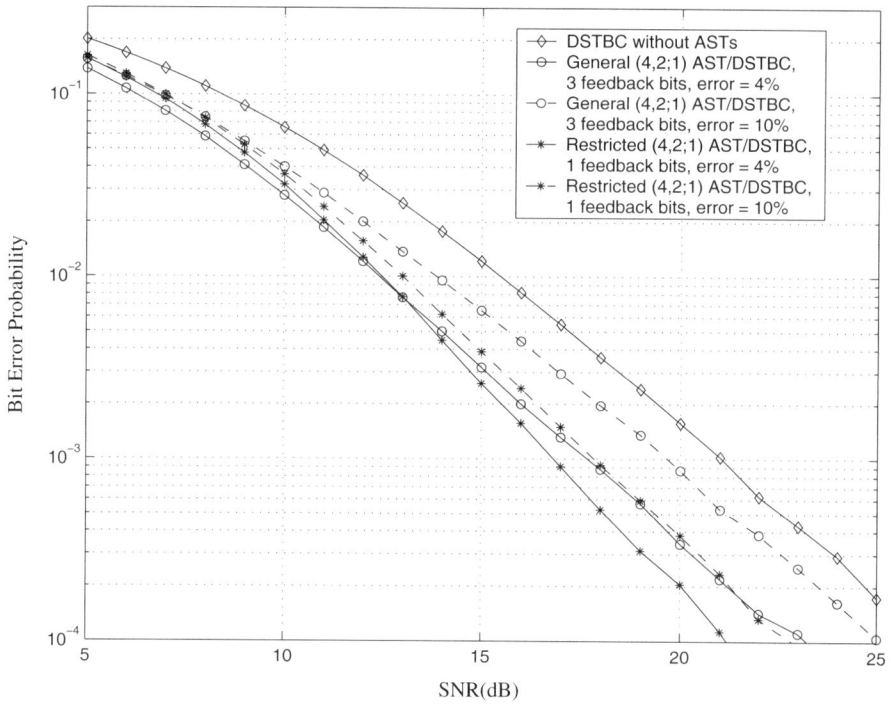

Figure 7.12. Proposed (4,2;1) AST/DSTBC schemes in correlated Rayleigh fading channels corresponding to the set $\mathcal{K}_A$ of covariance matrices.

compared to the case of the Alamouti DSTBC without AST/DSTBC schemes. Similarly, to achieve the same $BER = 10^{-3}$, the SNR gains achieved by the *restricted* (4,2;1) AST/DSTBC is 4.2 dB with the feedback error rate of 4% and 3.1 dB with the feedback error rate of 10%, compared to the case of the Alamouti DSTBC without AST/DSTBC schemes. It is interesting in this case that the *restricted* (4,2;1) AST/DSTBC provides better error performance than the *general* (4,2;1) AST/DSTBC at high SNRs. This is because the *restricted* (4,2;1) AST/DSTBC requires only 1 feedback bit while the *general* (4,2;1) AST/DSTBC requires 3 feedback bits. Therefore, for the same feedback error rate, the feedback information in the *restricted* (4,2;1) AST/DSTBC is less likely erroneous than that in the *general* (4,2;1) AST/DSTBC. As a result, in some special cases, the *restricted* schemes can provide better error performance than the *general* schemes. This observation has also been mentioned earlier in Section 6.3.7.

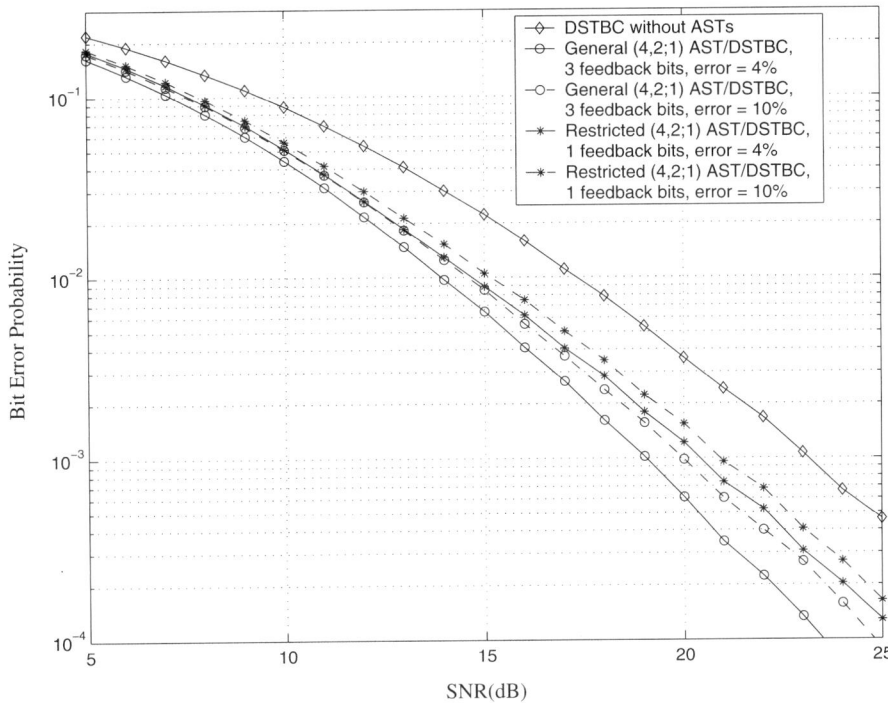

Figure 7.13. Proposed (4,2;1) AST/DSTBC schemes in correlated Rayleigh fading channels corresponding to the set $\mathcal{K}_B$ of covariance matrices.

Finally, Fig. 7.13 confirms further that, in the more correlated fading channels corresponding to the set $\mathcal{K}_B$, the superiority of our proposed ASTs still holds. To achieve the same $BER=10^{-3}$, the SNR gains achieved by the *general* (4,2;1) AST/DSTBC and the *restricted* (4,2;1) AST/DSTBC are 4 dB and 2.6 dB, respectively, with the feedback error rate of 4% and 3 dB and 2 dB, respectively, with the feedback error rate of 10%, compared to the case of the Alamouti DSTBC without our AST/DSTBC schemes.

7.4 Effect of Imperfect Carrier Recovery on the Performance of the Diversity Antenna Selection Techniques in Wireless Channels Utilizing DSTBCs

It has been shown in Section 6.3 that, the two proposed AST/DSTBC schemes, which are referred to as the *general* $(M, N; K)$ AST/DSTBC and the *restricted* $(M, N; K)$ AST/DSTBC, improve significantly the bit error performance of channels using DSTBCs. It has been further confirmed in Section 7.3 of this chapter that, the superiority of the proposed AST/DSTBC schemes still holds in the case of correlated Rayleigh flat fading channels. These conclusions are derived from an assumption that the carrier recovery at the receiver is perfect.

In reality, carrier phase/frequency recovery errors may exist resulting in degradation of the performance of the proposed AST/DSTBC schemes. Phase/frequency errors may be due to the difference between the frequency of the local oscillators in the transmitter and the receiver. They also may be caused by the Doppler frequency shift effect. Therefore, the effect of such errors on the performance of the proposed AST/DSTBC schemes should be considered.

In the rest of this chapter, we examine the effect of imperfect carrier phase/frequency recovery at the receiver on our AST/DSTBC schemes proposed for channels using DSTBCs. More specifically, the proposed AST/DSTBC schemes associated with the Alamouti DSTBC are considered. It is tedious to generalize the analysis mentioned here for other DSTBCs of higher orders, and therefore, we do not deal with this task in this book. In addition, we restrict ourselves to analyze the *general* $(M, N; K)$ AST/DSTBC schemes only. The *restricted* $(M, N; K)$ AST/DSTBC schemes can be similarly analyzed.

7.4.1   Effect of Phase Errors on the Performance of the Proposed Antenna Selection Techniques

Let T_c be the coherence time of the channel, i.e., the duration in which the transmission coefficients may be considered to be constant [Rappaport, 2002]. For ease of exposition, we assume that each coherence time T_c contains $\mathcal{L}$ symbol time slots (STSs). It should be emphasized that, by this assumption, we do not mean the channel is a slow Rayleigh fading one. Readers should refer to Example 6.3.2 in Section 6.3.1.2 for more details. In this example, each coherence time T_c of the channel contains 16 STSs, i.e. $\mathcal{L} = 16$. If the Alamouti code [Alamouti, 1998] is used, then the transmission coefficients can be considered to be unchanged during 8 consecutive code blocks. In this case, it is either impractical or uneconomical to use STBCs with *coherent detection* since the coherence time is too short to transmit multiple training symbols in order for the receiver to estimate the channel coefficients. In other words, the channel considered in the example can still be fast enough so that the utilization of DSTBCs (with *differential detection*) is useful.

Further, we assume that, due to imperfect carrier recovery, the *initial* phase error for each coherence time T_c is $\Delta\phi_0$ (radians). Meanwhile, *non-initial* phase errors are assumed to cumulate by adding a constant volume $\Delta\phi$ (radians) per STS to the phase of received signals. It means that the signal received in the kth STS, for $k = 1 \ldots \mathcal{L}$, in the considered coherence time T_c is now multiplied with $e^{i[(k-1)\Delta\phi + \Delta\phi_0]}$.

However, we realize that the bit error performance of DSTBCs does not depend on the initial phase error $\Delta\phi_0$. This is interpreted as follows. Like the differential phase shift keying (DPSK) demodulation, where the demodulation of signals only depends on the *relative phase difference* between consecutive signals, rather than the *absolute phase* of each signal, the bit error performance of DSTBCs does not depend on the initial phase error $\Delta\phi_0$, but depending on non-initial phase errors.

Therefore, in the rest of this chapter, we omit the initial phase error and only examine the effect of non-initial phase errors. If we denote Δf to be the frequency error per STS, it follows that

$$\Delta\phi = 2\pi \Delta f T_s = 2\pi \overline{\Delta f} \qquad (7.34)$$

where T_s is the period of each STS and $\overline{\Delta f} = \Delta f T_s$ is called the *equivalent frequency error*, which is the frequency error Δf normalized by the frequency of transmitted symbols $F_s = \frac{1}{T_s}$.

First, we examine the case of the Alamouti DSTBC *without* using our antenna selection schemes. A system containing 2 Tx antennas and 1 Rx antenna is considered.

As mentioned earlier in Section 6.3, in all conventional DSTBCs proposed in the literature, it is required that the transmission coefficients must be constant during at least two adjacent code blocks. For the Alamouti DSTBC, the transmission coefficients must be constant in a duration of at least 4 STSs where the initial, identity matrix $\mathbf{W}_0 = \mathbf{I}_2$ and, at least, one next code block $\mathbf{W}_1$ can be transmitted.

Without loss of generality, we merely consider a *typical* coherence time T_c. The analysis is similar for other coherence durations. Since the analysis is derived for a typical coherence time T_c, for simplicity, we omit the indices indicating the different coherence durations. We denote a_j and n_{tj}, for $j = 1, 2$ and $t = 0, 1, 2, etc.$, to be the transmission coefficients between the jth Tx antenna and the Rx antenna, and the noise affecting the receiver during the tth transmission (in the considered coherence time T_c), respectively. Note that the 0th transmission means the *initial* transmission.

The expressions given below are straightforwardly derived according to the DSTBCs proposed in [Ganesan and Stoica, 2002b]. Readers should refer to Section 6.3.1.1 of this book or [Ganesan and Stoica, 2002b] for more details. It is noted that the Alamouti DSTBC is considered here.

The transmission starts with an initial, identity matrix $\mathbf{W}_0 = \mathbf{I}_2$ carrying *no* information. The code block transmitted at time t, for $t = 1, 2, 3, etc.$, is given by (see Eq. (6.7) in Chapter 6)

$$\mathbf{W}_t = \mathbf{W}_{t-1}\mathbf{Z}_t \tag{7.35}$$

where $\mathbf{Z}_t$ is the Alamouti DSTBC defined as

$$\mathbf{Z}_t = \frac{1}{\sqrt{2}} \begin{bmatrix} s_1 & s_2 \\ -s_2^* & s_1^* \end{bmatrix} \tag{7.36}$$

Each pair of transmitted symbols is taken once at a time from the input data stream to generate $\mathbf{Z}_t$. Therefore, both transmitted symbols in (7.36) should be indexed by the index t, i.e. $s_1^{(t)}$ and $s_2^{(t)}$ for instance. However, for short, we omit the index t of the transmitted symbols in the expression of $\mathbf{Z}_t$.

Let us calculate the matrix $\mathbf{W}_1$ for instance. We have

$$\mathbf{W}_1 = \mathbf{W}_0\mathbf{Z}_1 = \mathbf{I}_2\mathbf{Z}_1 = \mathbf{Z}_1 = \frac{1}{\sqrt{2}} \begin{bmatrix} s_1 & s_2 \\ -s_2^* & s_1^* \end{bmatrix} \tag{7.37}$$

The received matrix at time t is (see Eq. (6.8) in Chapter 6)

$$\mathbf{R}_t = \mathbf{A}\mathbf{W}_t + \mathbf{N}_t \tag{7.38}$$

Following (7.38), the signals received during the initial transmission when $\mathbf{W}_0 = \mathbf{I}_2$ is transmitted are calculated as (the first subscript of the received signals refers to the index t while the second subscript indicates the order of the corresponding STS)

$$
\begin{aligned}
r_{01} &= a_1 + n_{01} , \\
r_{02} &= a_2 e^{i\Delta\phi} + n_{02}
\end{aligned}
$$

Following (7.37) and (7.38), the signals received when the second code block $\mathbf{W}_1 = \mathbf{Z}_1$ is transmitted are calculated as

$$r_{11} = \frac{1}{\sqrt{2}}(a_1 s_1 + a_2 s_2)e^{i2\Delta\phi} + n_{11} , \tag{7.39}$$

$$r_{12} = \frac{1}{\sqrt{2}}(-a_1 s_2^* + a_2 s_1^*)e^{i3\Delta\phi} + n_{12} \tag{7.40}$$

The received signals corresponding to the transmitted code blocks $\mathbf{W}_2$, $\mathbf{W}_3$, etc. are similarly analyzed, based on Eq. (7.35) and Eq. (7.38). The above expressions of the received signals are used in the simulation in the next section.

Next, we consider the Alamouti DSTBC in association with our *general* (3,2;1) AST/DSTBC and *general* (4,2;1) AST/DSTBC schemes. The *general* (4,2;1) AST/DSTBC scheme is analyzed here in detail for illustration. The *general* (3,2;1) AST/DSTBC scheme can be similarly analyzed. The expressions given below are straightforwardly derived according to our proposed AST/DSTBC schemes. Readers should refer to Section 6.3.4 for more details.

Like all conventional DSTBC techniques, transmission coefficients in the Alamouti DSTBC associated with our *general* (4,2;1) AST/DSTBC are also required to be constant during at least two consecutive code blocks. This means that transmission coefficients must be constant during the channel coherence time T_c of at least 6 STSs, which are corresponding to the time when the initial block $\tilde{\mathbf{W}}_0 = \mathbf{I}_4$ and, at least, one next block $\mathbf{W}_1$ are transmitted.

From (7.35) and (7.38), we have the received signals during the *initial* transmission as given below

$$
\begin{aligned}
r_{01} &= a_1 + n_{01} \\
r_{02} &= a_2 e^{i\Delta\phi} + n_{02} \\
r_{03} &= a_3 e^{i2\Delta\phi} + n_{03} \\
r_{04} &= a_4 e^{i3\Delta\phi} + n_{04}
\end{aligned}
$$

Following the proposed *general* (4,2;1) AST/DSTBC scheme, in each channel coherence time T_c, the best two Tx antennas (among the four available Tx antennas) are selected to transmit the rest of data during the considered coherence time. These two Tx antennas provide the first two largest instantaneous powers of the signals received during the initial transmission in the considered coherence time. Without loss of generality, we assume that, during the current channel coherence time, the best 2 Tx antennas are the 1st and 2nd ones. It means that the code blocks after the initial block $\tilde{\mathbf{W}}_0$ are transmitted via the 1st and 2nd Tx antennas.

As analyzed in Remark 6.3.5 of Section 6.3.4, after the initial block $\tilde{\mathbf{W}}_0$ is transmitted, the next code blocks $\mathbf{W}_1$, $\mathbf{W}_2$, etc. are calculated following (7.35), by using a *tacit* default, identity matrix $\mathbf{W}_0 = \mathbf{I}_2$.

Let us consider the second code block $\mathbf{W}_1$ for example. This code block is generated by multiplying the matrix $\mathbf{Z}_1$ in (7.36) with a tacit, default, identity matrix $\mathbf{W}_0 = \mathbf{I}_2$, and consequently, $\mathbf{W}_1$ has the same expression as (7.37), i.e., $\mathbf{W}_1 = \mathbf{Z}_1$.

The received signals are still computed following (7.38). Therefore, when the code block $\mathbf{W}_1$ is transmitted, we have the following received signals:

$$r_{11} = \frac{1}{\sqrt{2}}(a_1 s_1 + a_2 s_2)e^{i4\Delta\phi} + n_{11}, \qquad (7.41)$$

$$r_{12} = \frac{1}{\sqrt{2}}(-a_1 s_2^* + a_2 s_1^*)e^{i5\Delta\phi} + n_{12} \qquad (7.42)$$

The received signals corresponding to the transmitted code blocks W_2, W_3, etc. are similarly analyzed, based on Eq. (7.35) and Eq. (7.38). The aforementioned expressions of the received signals are used in the simulation in the next section.

7.4.2   Simulation Results

In this section, bit error rates (BER) of the Alamouti DSTBC *with* and *without* our proposed AST/DSTBC schemes versus the ratio between bit energy and noise energy (E_b/N_0) are presented. The channel is assumed to be an independent, flat Rayleigh fading one. The feedback error rates of 4% and 10% are considered in simulations when the Alamouti DSTBC is associated with our proposed AST/DSTBC schemes.

The bit error performance of the Alamouti DSTBC *without* antenna selection is shown in Fig. 7.14. Again, the transmission coefficients must remain unchanged during two adjacent code blocks. Hence, this simulation is run with the transmission coefficients being constant during *at least* 4 STSs (2 STSs for the initial block $\mathbf{W}_0$ and 2 STSs for the next code block $\mathbf{W}_1$).

The performance of the Alamouti DSTBC *with* our proposed *general* (3,2;1) AST/DSTBC is presented in Fig. 7.15. This simulation is run for the channel coherence time T_c of at least two consecutive code blocks, i.e., of *at least* 5 STSs (3 STSs for the initial code block $\tilde{\mathbf{W}}_0 = \mathbf{I}_3$ and 2 STSs for the next code block $\mathbf{W}_1$).

The performance of the Alamouti DSTBC *with* our proposed *general* (4,2;1) AST/DSTBC is presented in Fig. 7.16. Similarly, the simulation is run for the channel coherence time T_c of *at least* 6 STSs (4 STSs for the initial code block $\tilde{\mathbf{W}}_0 = \mathbf{I}_4$ and 2 STSs for the next code block $\mathbf{W}_1$).

In all simulations, a unitary QPSK signal constellation, where the power of the symbols in the constellation is one, is considered. The transmission coefficients and noise are assumed to be i.i.d. complex Gaussian random variables with the distributions $\mathcal{CN}(0, 1)$ and $\mathcal{CN}(0, \sigma^2)$, respectively. From (7.39), (7.40), (7.41) and (7.42) and the aforementioned assumptions, we realize that the received power for each symbol during each STS is 1/2, and consequently, the energy per bit is $E_b = 1/4$. Therefore, in all simulations, we assign the

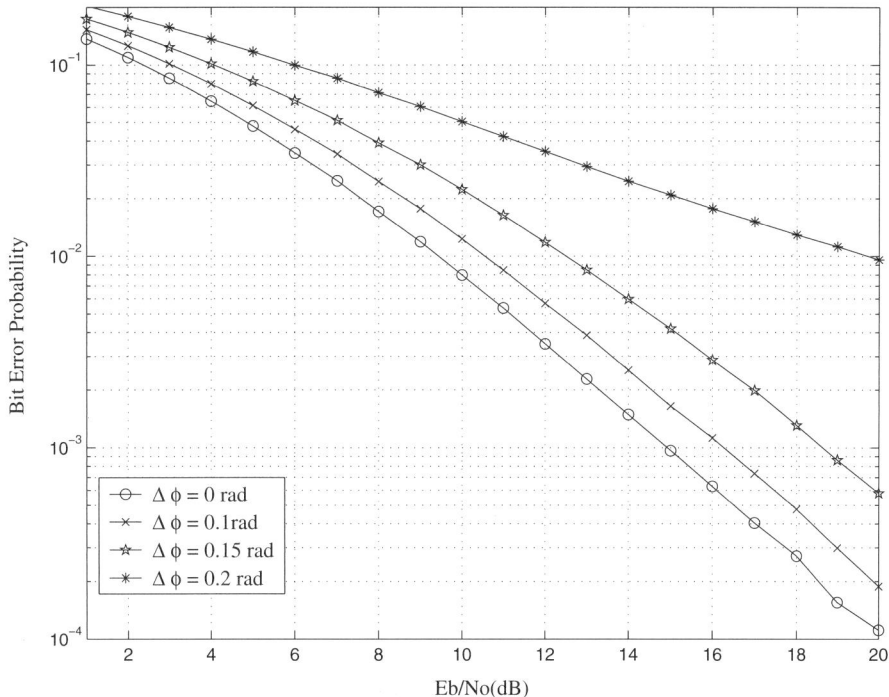

Figure 7.14.   The effect of imperfect phase recovery on the performance of the Alamouti DSTBC without our ASTs.

noise variance (or noise power) to $\sigma^2 = \frac{1}{4(E_b/N_0)}$ in order for E_b/N_0 to be the channel SNR per bit.

We can see from Fig. 7.14 that, when the proposed AST/DSTBC schemes are *not* utilized and the phase errors $\Delta\phi = 0.1$ and $\Delta\phi = 0.15$ rad exist, the E_b/N_0 required to achieve the same $BER = 10^{-3}$ as in the case *without* phase errors ($\Delta\phi = 0$) is approximately 1.3 and 3.75 dB higher, respectively. The bit error performance of the Alamouti DSTBC degrades rapidly for $\Delta\phi > 0.15$ rad.

Likewise, in our *general* (3,2;1) AST/DSTBC scheme, the E_b/N_0 should be 1 dB and 2.6 dB higher, corresponding to the phase errors of 0.05 rad and 0.075 rad, respectively (the feedback error rate can be either 4% or 10%), to achieve the same $BER = 10^{-3}$ as in the case *without* phase errors ($\Delta\phi = 0$). The phase errors being greater than 0.075 rad seriously degrade the performance of the *general* (3,2;1) AST/DSTBC scheme.

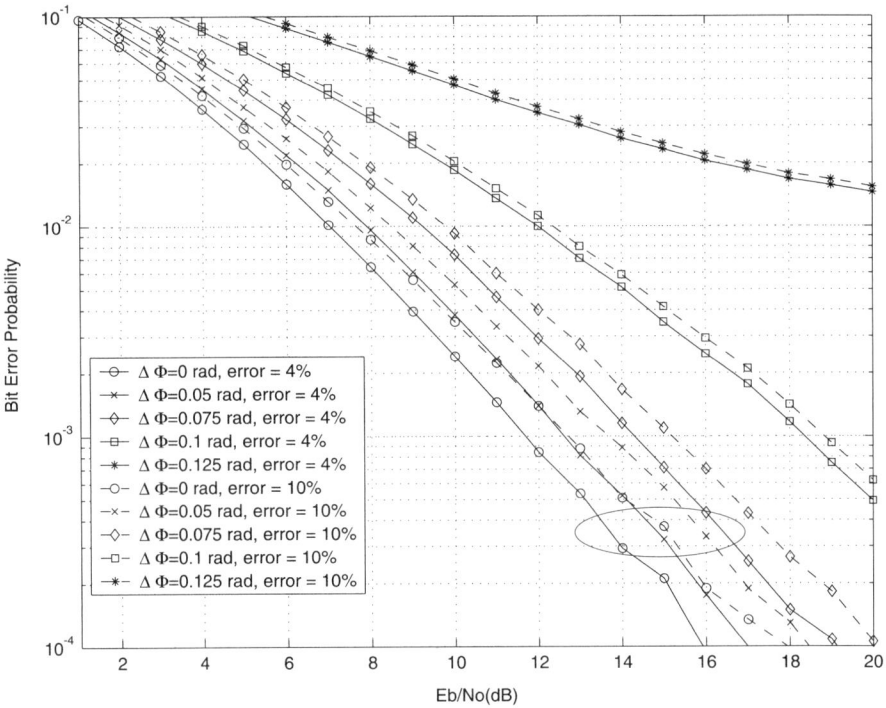

Figure 7.15.   The effect of imperfect phase recovery on the performance of the general (3,2;1) AST/DSTBC scheme.

Similarly, in the *general* (4,2;1) AST/DSTBC scheme, the E_b/N_0 should be 0.75 dB and 2 dB higher to achieve the same $BER = 10^{-3}$ as in the case *without* phase errors, corresponding to the phase errors of 0.05 rad and 0.075 rad, respectively, in both cases where the feedback error rate is either 4% or 10%. The phase errors being greater than 0.075 rad also seriously degrade the performance of the *general* (4,2;1) AST/DSTBC scheme.

We may have an observation that DSTBCs associated with the proposed AST/DSTBC schemes are more sensitive to phase errors than those codes without the AST/DSTBC schemes. From the simulation results, we also may deduce that the tolerance of the Alamouti DSTBC to phase errors in the cases *with* and *without* the *general* (3,2;1) AST/DSTBC (as well as the *general* (4,2;1) AST/DSTBC) is 0.075 rad (or 13.5^o) and 0.15 rad (or 27°), respectively.

In addition, Figures 7.14, 7.15 and 7.16 show that, when the Alamouti DSTBC is associated with the *general* (3,2;1) AST/DSTBC scheme and the

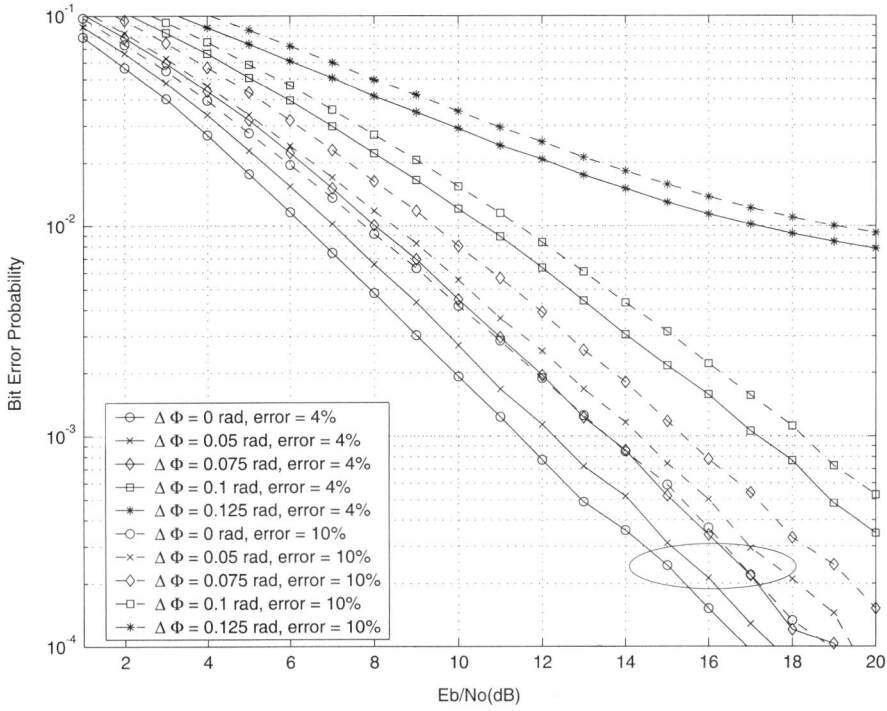

Figure 7.16. The effect of imperfect phase recovery on the performance of the general (4,2;1) AST/DSTBC scheme.

feedback error rate is 4%, the Alamouti DSTBC has better bit error performance than the Alamouti DSTBC *without* antenna selection (*no matter* whether phase errors exist or not, i.e., corresponding to all curves in Fig. 7.14), provided that the phase errors satisfy $\Delta\phi \leq 0.075$ rad. When the feedback error rate is 10%, this superiority of the *general* (3,2;1) AST/DSTBC scheme still holds, provided that $\Delta\phi \leq 0.05$ rad. The curves corresponding to $\Delta\phi \leq 0.075$ rad (when the feedback error rate is 4%) and $\Delta\phi \leq 0.05$ rad (when the feedback error rate is 10%) are grouped in Fig. 7.15. These observations are also true for the Alamouti DSTBC associated with the *general* (4,2;1) AST/DSTBC scheme. The curves corresponding to $\Delta\phi \leq 0.075$ rad (when the feedback error rate is 4%) and $\Delta\phi \leq 0.05$ rad (when the feedback error rate is 10%) are grouped in Fig. 7.16.

Therefore, it might be stated that the Alamouti DSTBC associated with the proposed *general* (3,2;1) AST/DSTBC and *general* (4,2;1) AST/DSTBC overwhelms the Alamouti DSTBC without antenna selection, even when some certain values of phase errors exist.

It is worth to note that Figures 7.14, 7.15 and 7.16 themselves also present the BER versus the E_b/N_0 in the case of frequency recovery errors, since, frequency errors and phase errors follow the relation (7.34). It means that the tolerance to equivalent frequency errors for the Alamouti DSTBC *with* and *without* utilizing antenna selection is $\overline{\Delta f} = \frac{0.075}{2\pi} = 1.19 \times 10^{-2}$ and $\overline{\Delta f} = \frac{0.15}{2\pi} = 2.39 \times 10^{-2}$, respectively. In other words, the tolerance to frequency errors in the case of the Alamouti DSTBC *with* and *without* our AST/DSTBC shemes is approximately 1.19 % and 2.39 % of the symbol frequency $F_s = \frac{1}{T_s}$, respectively.

7.5   Conclusions

In this chapter, we have derived a more generalized algorithm to generate correlated Rayleigh fading envelopes. Using the presented algorithm, one can generate an arbitrary number N of *either* Rayleigh envelopes with any desired power $\sigma_{r_j}^2$, $j = 1, \ldots, N$, *or* those envelopes corresponding to any desired power $\sigma_{g_j}^2$ of Gaussian random variables. This algorithm also facilitates to generate *equal* as well as *unequal* power Rayleigh envelopes. It is applicable to both scenarios of *spatial correlation* and *spectral correlation* between the random processes. The coloring matrix is determined by a positive semi-definiteness forcing procedure and an eigen decomposition procedure without using Cholesky decomposition. Consequently, the restriction on the positive definiteness of the covariance matrix is relaxed and the algorithm works well without being impeded by the roundoff errors of MatLab. The proposed algorithm can be used to generate Rayleigh envelopes corresponding to any desired covariance matrix, no matter whether or not it is positive definite. In comparison with the conventional methods, besides being *more generalized*, our proposed algorithm (with or without Doppler spectrum spread) is *more precise*, while overcoming all shortcomings of the conventional methods.

Based on the proposed algorithm, this chapter then examines our AST/DSTBC schemes proposed for wireless systems utilizing DSTBCs in spatially correlated flat Rayleigh fading channels. This correlation may be due to insufficient spatial separation between Tx/Rx antennas. As mentioned in Chapter 6 of this book and in [Tran and Wysocki, 2003], [Tran and Wysocki, 2004], DSTBCs combined with the proposed AST/DSTBC schemes have much better performance than those without antenna selection. It is further concluded that the proposed AST/DSTBC schemes not only work well in *uncorrelated* fading channels, but they are also very robust in *correlated* channels.

Finally, the chapter examines the effect of imperfect carrier phase/frequency recovery at the receiver on our AST/DSTBC schemes proposed for channels using DSTBCs. Phase/frequency errors may be due to the difference between the frequency of the local oscillators in the transmitter and the receiver. They also may be caused by the Doppler frequency shift effect, such as in the rapid Rayleigh fading environment. Simulation results provide further conclusions that, although DSTBCs associated with the proposed AST/DSTBC schemes are more sensitive to the imperfect carrier recovery, for certain phase errors, the DSTBCs still have better bit error performance than those codes without antenna selection (no matter whether phase errors exist or not). In addition, the tolerance to phase errors and frequency errors in the cases of DSTBCs with and without using antenna selection has also been derived in this chapter.

Chapter 8

CONCLUSIONS

8.1 Introduction

In this chapter, we bring together the main conclusions reported in the previous chapters. We also provide some recommendations on the utilization of Complex Orthogonal Space-Time Block Codes (CO STBCs) and the proposed Antenna Selection Techniques (ASTs) in wireless channels. Finally, we list some important issues for the future work in these areas.

8.2 Main Conclusions

Novel constructions of improved, square CO STBCs

The improved, square CO STBCs are referred to as the square CO STBCs with fewer or even with no zero entries, where the transmitted symbols tend to be equally dispersed through transmitter (Tx) antennas as well as through symbol time slots (STSs). The benefits of the improved CO STBCs are evident as analyzed in Chapters 3, 4 and 5. In this book, first, we have derived three new, order-8, *maximum rate* CO STBCs with a few or even without zeros, which are superior to the conventional order-8 CO STBCs.

Later, by modifying the Williamson and Wallis-Whiteman arrays to apply to complex matrices, we have proposed two *new* methods of constructing *square*, order-$4n$ CO STBCs from square, order-n CO STBCs which satisfy certain properties. Although constructions of square, maximum rate CO STBCs are well known [Liang, 2003], the codes resulting from these constructions have numerous zeros, since these construction methods always involve identity matrices. This disadvantage is overcome by our proposed constructions, where the sub-matrices of order n are wisely selected to include as few zeros as possible or even no zeros. A smaller number of zeros in the sub-matrices results in a smaller number of zeros in the resultant CO STBC. Our methods also partially overcome the difficulty in designing codes from Amicable Orthogonal Designs (AODs) since our methods mainly require to find two $n \times n$ matrices satisfying our conditions, as opposed to requiring to find a number of $4n \times 4n$ weighting matrices as in the AOD constructions.

Applying the proposed constructions, we have constructed various *square*, *maximum rate*, order-8 CO STBCs with no zeros. In our CO STBCs, the transmitted symbols disperse equally through Tx antennas, which results in an equal transmitted power via each Tx antenna during every STS.

Multi-modulation schemes to increase data rates and the optimal inter-symbol power allocation

A number of the *square*, *maximum rate* CO STBCs proposed in the literature, such as in [Tarokh et al., 1999b] and in this book, have a property that some transmitted symbols appear more often than the others. This property is exploited to increase the data rates by using Multi-Modulation Schemes (MMSs), where a high level modulation scheme is applied to the symbols appearing more often than the others. The higher level modulated symbols are transmitted with more power to carry more information bits.

We would like to emphasize that the maximum rates of CO STBCs bounded by Eq. (2.31) and Eq. (2.33) in Chapter 2, corresponding to *square* and *non-square* constructions respectively, are the upper bounds for the code rates of CO STBCs *without* the consideration of modulation. When MMSs are considered, the code rates have *no* upper bounds. Instead, the code rates of CO STBCs in this case depend on the modulation constellations. For the M-PSK or M-QAM constellation, the rate would be different for different M.

Certainly, when MMSs are employed, there usually exists a tradeoff between the code rate and the bit error performance in the sense that a higher code rate results in a worse bit error performance. However, the important result mentioned in the book is that, for the same MMSs, the new CO STBCs proposed in this book may provide better error performance, compared to the conventional, same order CO STBCs. It is interesting that the increase in the transmitted power of higher level modulated symbols in the proposed MMSs does not impede, but facilitates the practical implementation. This is due to the fact that the increase in the power of higher level modulated symbols results in an equal power transmission through each Tx antenna during each STS.

In addition, the optimal inter-symbol power allocation in MMSs to achieve the probably best error performance should be examined. The book thus proposes a method to evaluate the (smallest) optimal power ratio between the higher level modulated symbols and the lower level modulated ones. Although, the proposed method is mentioned here for order-8 CO STBCs, it can be applied to CO STBCs of any order, provided that some transmitted symbols in the code blocks appear more than once in each column (also in each row). The method proposed in the book for evaluating the optimal inter-symbol power allocation is more practical than the conventional method [Tirkkonen and Hottinen, 2001]. Particularly, in our method, we search for the *'globally'* optimal value, rather than the *'locally'* optimal value as in [Tirkkonen and Hottinen, 2001].

Improved diversity antenna selection technique for channels using Space-Time Block Codes (STBCs) with coherent detection

Motivated by the N-out-of-M AST proposed in [Katz et al., 2001], in this book, we have proposed a simple, improved, closed loop AST, which is referred to as the $(N+1, N; K)$ AST/STBC scheme, to improve further the performance of wireless channels using STBCs with *coherent detection*. Our technique provides the same bit error performance as the N-out-of-$(N+1)$ AST proposed in [Katz et al., 2001] if the receivers in both techniques have K Rx antennas. However, the $(N + 1, N; K)$ AST/STBC requires shorter time to process the feedback information due to our proposed structure of the feedback information. The processing time benefit is especially considerable when the proposed $(N + 1, N; K)$ AST/STBC scheme is combined with the Alamouti show that the proposed $(N + 1, N; K)$ AST/STBC scheme improves AST/STBC scheme improves significantly performance of the channels using STBCs.

Diversity antenna selection techniques for channels using Differential Space-Time Block Codes (DSTBCs) with differential detection

As opposed to the diversity ASTs proposed for channels utilizing STBCs with *coherent detection*, ASTs for channels utilizing DSTBCs with *differential detection* have not yet been intensively considered. In the book, we have proposed two ASTs referred to as the *general* $(M, N; K)$ AST/DSTBC scheme, and the *restricted* $(M, N; K)$ AST/DSTBC scheme for channels using DSTBCs with arbitrary numbers of Tx and Rx antennas. In the proposed AST/DSTBC schemes, the initial code block transmitted at the beginning of each channel coherence time is not only used to initialize the transmission, but also used to predict semi-blindly the channel properties. This prediction is, in turn, used for diversity antenna selection purposes. Therefore, the good transmitter antennas can be selected with a limited number (typically, 1 or 2) of training symbols. Simulation results show that the proposed AST/DSTBC schemes improve significantly the bit error performance of wireless channels using DSTBCs.

Generalized algorithm for generating correlated Rayleigh fading envelopes

Motivated by the examination of the proposed AST/DSTBC schemes in correlated Rayleigh fading channels, we realize that although the methods for generating correlated Rayleigh fading envelopes have been intensively considered in the literature, those methods are not complete and have their own shortcomings which significantly limit their applicability. Hence, we have derived a *more generalized* algorithm, which can be used to generate *any number* of Rayleigh fading envelopes with *any power* (*equal or unequal* power) corresponding to *any desired* covariance matrix and is applicable to both *discrete-time instant* scenario and *real-time* scenario. The algorithm also overcomes the roundoff errors, which may cause the interruption of Matlab programs, due to the fact that a positive semi-definiteness forcing procedure is used in this algorithm and the coloring matrix of the covariance matrix is determined by eigen decomposition, rather than Cholesky decomposition. The algorithm can work well, regardless of the positive definiteness of the covariance matrix. Further, the algorithm provides a straightforward method for the generation of complex Gaussian random variables (with Rayleigh envelopes) with correlation properties as functions of *time delay* and *frequency separation* (as in Orthogonal Frequency Division Multiplexing (OFDM) systems), or *spatial separation* between transmission antennas like with multiple antennas in Multiple Input Multiple Output (MIMO) systems. The applicability of the algorithm is therefore much broadened, compared to all conventional methods. Apart from being *more generalized*, the proposed algorithm is also *more accurate*, while overcoming all shortcomings of the conventional methods.

Robustness of the proposed AST/DSTBC schemes in correlated Rayleigh fading channels

Based on the proposed algorithm for generating correlated Rayleigh fading envelopes, we have examined the performance of the diversity AST/DSTBC schemes proposed for channels using DSTBCs in the correlated Rayleigh fading scenario. The correlation in the channels may be due to the insufficient spatial separation between the Tx and/or Rx antennas when the AST/DSTBC schemes are utilized in MIMO systems, or due to the spectral correlation between different carrier frequencies when they are utilized in OFDM systems. Simulation results show that the proposed AST/DSTBC schemes not only work well in *independent* Rayleigh fading channels, but also are very robust in *correlated* Rayleigh fading channels.

Performance of the proposed AST/DSTBC schemes in the imperfect carrier recovery scenario

Finally, the book examines the effect of imperfect carrier phase/frequency recovery at the receiver on our AST/DSTBC schemes proposed for channels using DSTBCs. Phase/frequency errors may be due to the Doppler frequency shift effect, such as in a rapid Rayleigh fading environment. They also may be caused by the difference between the frequency of the local oscillators in the transmitter and the receiver. Simulation results provide further conclusions that, although DSTBCs associated with the proposed AST/DSTBC schemes are more sensitive to the imperfect carrier recovery, but for certain phase errors, the DSTBCs still have better bit error performance than those without antenna selection (no matter whether phase errors exist or not). In addition, the tolerance to phase errors and frequency errors in the cases of DSTBCs with and without using antenna selection has also been derived. The examination of the performance of the proposed AST/DSTBC schemes in the imperfect carrier recovery scenario confirms further the robustness of DSTBCs in association with the proposed AST/DSTBC schemes over those codes without antenna selection.

8.3 Recommendations

Based on the works presented in this book, we can derive the following recommendations:

1 From the practical point of view, besides the rank and determinant criteria [Tarokh et al., 1999c], [Tarokh et al., 1998], the two following criteria should be considered in designing a good CO STBC, especially in a wireless system of high data rates. First, the CO STBC should contain *as few zero entries as possible*. Second, a better dispersion of the transmitted symbols in both spatial (between the Tx and/or Rx antennas) and temporal directions more likely (but not necessarily) results in a better bit error performance. Mathematically, each indeterminate in the code matrix should be *as much scattered in the whole matrix as possible*.

2 To evaluate the optimal inter-symbol power allocation in MMSs, the method proposed in this book should be used, rather than the method proposed in [Tirkkonen and Hottinen, 2001] which has a marginal meaning from the practical point of view.

3 When the generalized Williamson and Wallis-Whiteman constructions are used to construct *square* CO STBCs, sub-matrices should be wisely selected to include as few zeros as possible or even no zeros, while they must satisfy the certain properties as mentioned in the proposed constructions. A smaller number of zeros in the sub-matrices results in a smaller number of zeros in the resultant CO STBC. Consequently, the resultant CO STBC will contain no zero entries if all sub-matrices contain no zero entries.

4 To generate correlated Rayleigh fading envelopes with *the presence of the Doppler frequency shift effect*, the generalized algorithm derived in the book should be used, rather than the method proposed in [Sorooshyari and Daut, 2003] which results an *incorrect* covariance matrix of the resultant Rayleigh envelopes.

5 In reality, carrier recovery at the receiver can be imperfect, which may be characterized by a certain frequency error. In such a case, the frequency F_s of the transmitted symbols in the Alamouti DSTBC should be large enough so that the frequency error is *at most* equal to 1.19 % (if the Alamouti DSTBC is associated with our proposed AST/DSTBC schemes) and 2.39 % (if the Alamouti DSTBC is utilized without our AST/DSTBC schemes) of the symbol frequency F_s, respectively. Otherwise, the bit error performance of the differential space-time coded channels (with or without our proposed AST/DSTBC schemes) decreases significantly.

8.4   Future Works

It has been known that relaxing the definition of CO STBCs to provide linear processing (LP) at the transmitter fails to increase the code rate of *square* CO STBCs [Su and Xia, 2003], [Tarokh et al., 1999b] if the condition $l_{j1} = \cdots = l_{jk}$ (see Section 2.3.2.5 for more details) is satisfied. However, it is *unknown* whether the maximum rate of *square* CO STBCs with LP is different to that of *square* CO STBCs without LP if the condition $l_{j1} = \cdots = l_{jk}$ is *not* satisfied. We have also concluded in Chapter 2 that the *known, maximum rate* of *non-square* CO STBCs *with LP* to date is the same as that of *non-square* CO STBCs *without LP* for any number of Tx antennas. However, it is *unknown* whether the *true* maximum rate of *non-square* CO STBCs *with LP* is higher than that of *non-square* CO STBCs *without LP*. These issues require to be further examined.

Our constructions based on the generalized Williamson and Wallis-Whiteman arrays may be used to design *square* CO STBCs of order 16 or 32 from *square* CO STBCs of order 4 or 8, respectively, provided that there exist the sub-matrices satisfying the conditions of our theorems. The constructions of *square* CO STBCs of higher orders, such as 16 or 32, require further study, and this is our future work.

The Kharaghani array [Kharaghani, 2000] can be modified to apply to complex symbols, and therefore, can be used to construct a *square*, order-$8n$ CO STBC from the order-n sub-matrices which satisfy certain properties. For instance, it may be possible to utilize the modified Kharaghani array to construct a *square*, order-16 CO STBC from several sub-matrices of order 2. Hence, generalization of the Kharaghani array to apply to complex symbols is the future work.

In this book, we have proposed some ASTs for channels using STBCs and DSTBCs. The future work would be the consideration of the proposed ASTs in association with Space-Time Turbo Codes or Space-Time Trellis Codes (STTCs), which provide coding gains unlike STBCs or DSTBCs, in either flat Rayleigh or frequency selective Rayleigh fading channels.

The model for examining the AST/DSTBC schemes, which are proposed for differential space-time coded wireless channels, in the scenario of imperfect carrier phase/frequency recovery is derived in this book. The performance as well as the sensitivity of the proposed AST/DSTBC schemes to the carrier recovery errors are also examined. It would be better and more complete if the solutions to mitigate the effect of the imperfect carrier recovery on the performance of the proposed AST/DSTBC schemes are derived. This should be further examined.

Glossary

3G   Third Generation wireless technology

3GPP   The Third Generation Partnership Project

AOD   Amicable Orthogonal Design

AST   Antenna Selection Technique

AWGN   Additive White Gaussian Noise

BER   Bit Error Rate

BLAST   Bell Lab Layered Space-Time

BPSK   Binary Phase Shift Keying

BS   Base Station

CDMA   Code Division Multiple Access

CO STBC   Complex Orthogonal Space-Time Block Code

COD   Complex Orthogonal Design

const   constant

DPCCH   Dedicated Physical Control Channel

DPSK   Differential Phase Shift Keying

DS-SS   Direct Sequence Spread Spectrum

DSTBC   Differential Space-Time Block Code

DSTM   Differential Space-Time Modulation

e.g.   exempli gratia

ECK   Exact Channel Knowledge

EGC   Equal Gain Combining

etc.   et cetera

FEC   Forward Error Correction

FH-SS   Frequency Hoping Spread Spectrum

GCOD   Generalized Complex Orthogonal Design

GSM   Global System for Mobile Communications

i.e   id est

i.i.d.   identically independently distributed

IDFT   Inverse Discrete Fourier Transform

iff   if and only if

ISI   Inter-Symbol Interference

LAN   Local Area Network

LOS   Line Of Sight

LST   Layered Space-Time Code

M-ary   Multiple Level Modulation

MC-SS   Multi-Carrier Spread Spectrum

MIMO   Multiple Input Multiple Output

MMS   Multi-Modulation Scheme

M-PSK   M-ary Phase Shift Keying

MRC   Maximum Ratio Combining

MS   Mobile Station

OFDM   Orthogonal Frequency Division Multiplexing

PAM   Pulse Amplitude Modulation

PCU   Per Channel Use

PDF   Probability Density Function

PSK   Phase Shift Keying

QAM   Quadrature Amplitude Modulation

QPSK   Quadrature Phase Shift Keying

rms   root-mean-square

Rx antenna   Receiver antenna

SC   Scanning Combining

SCK   Statistical Channel Knowledge

SER   Symbol Error Rate

SNR   Signal-to-Noise Ratio

STBC   Space-Time Block Code

STC Space-Time Code

STS Symbol Time Slot

STTC Space-Time Trellis Code

Tx antenna Transmitter antenna

w. r. t. with respect to

WCDMA Wideband Code Division Multiple Access

Appendix A
Symbol Error Probability of M-ary PSK Signals

In this section, we derive the approximated symbol error probability of M-ary PSK signals in *flat Rayleigh fading channels* when SNR per symbol is large enough. The symbol error probability of M-ary PSK signals in L-path Rayleigh fading channels is given below (see equation (14.4-38) in [Proakis, 2001]):

$$
\begin{aligned}
P_M &= \frac{(-1)^{L-1}(1-\mu^2)^L}{\pi(L-1)!} \times \left(\frac{\partial^{L-1}}{\partial b^{L-1}} \left\{ \frac{1}{1-\mu^2} \left[\frac{\pi(M-1)}{M} - \right. \right. \right. \\
&\quad \left. \left. \left. - \frac{\mu \sin(\frac{\pi}{M})}{\sqrt{1-\mu^2 \cos^2(\frac{\pi}{M})}} \cot^{-1}\left(\frac{-\mu \cos(\frac{\pi}{M})}{\sqrt{1-\mu^2 \cos^2(\frac{\pi}{M})}} \right) \right] \right\} \right)_{b=1}
\end{aligned}
\tag{A.1}
$$

where, by definition:

$$
\mu = \sqrt{\frac{\bar{\gamma}_c}{\bar{\gamma}_c + 1}} = \sqrt{\frac{(\bar{\gamma}_b \log_2 M)/L}{(\bar{\gamma}_b \log_2 M)/L + 1}}
\tag{A.2}
$$

and $\bar{\gamma}_c$ and $\bar{\gamma}_b$ are the average SNR per channel and per bit, respectively. In flat Rayleigh fading scenario, we have $L = 1$. Note that:

$$
\cot^{-1}\left(-\frac{\mu \cos(\frac{\pi}{M})}{\sqrt{1-\mu^2 \cos^2(\frac{\pi}{M})}} \right) = \pi - \cot^{-1}\left(\frac{\mu \cos(\frac{\pi}{M})}{\sqrt{1-\mu^2 \cos^2(\frac{\pi}{M})}} \right)
$$

then we have

$$
\begin{aligned}
P_M &= \frac{(1-\mu^2)}{\pi} \left\{ \frac{1}{1-\mu^2} \left[\frac{\pi(M-1)}{M} - \frac{\pi\mu \sin(\frac{\pi}{M})}{\sqrt{1-\mu^2 \cos^2(\frac{\pi}{M})}} \right. \right. \\
&\quad \left. \left. + \frac{\mu \sin(\frac{\pi}{M})}{\sqrt{1-\mu^2 \cos^2(\frac{\pi}{M})}} \cot^{-1}\left(\frac{\mu \cos(\frac{\pi}{M})}{\sqrt{1-\mu^2 \cos^2(\frac{\pi}{M})}} \right) \right] \right\}
\end{aligned}
$$

When the SNR per symbol satisfies: $\bar{\gamma}_c \gg 1$, such as $\bar{\gamma}_c \geq 10$ (i.e., 10 dB), then $\mu \approx 1$. Therefore, we have:

$$
\begin{aligned}
P_M &\approx \frac{(1-\mu^2)}{\pi}\left\{\frac{1}{1-\mu^2}\left[\frac{\pi(M-1)}{M} - \frac{\pi\mu\sin(\frac{\pi}{M})}{\sqrt{1-\mu^2\cos^2(\frac{\pi}{M})}}\right.\right.\\
&\quad + \left.\left.\frac{\mu\sin(\frac{\pi}{M})}{\sqrt{1-\mu^2\cos^2(\frac{\pi}{M})}}\cot^{-1}\left(\cot\left(\frac{\pi}{M}\right)\right)\right]\right\}\\
&= \frac{(M-1)(1-\mu^2)}{M}\left\{\frac{1}{1-\mu^2}\left[1 - \frac{\mu\sin(\frac{\pi}{M})}{\sqrt{1-\mu^2\cos^2(\frac{\pi}{M})}}\right]\right\}
\end{aligned}
$$

(A.3)

$$
= \frac{(M-1)(1-\mu^2)}{M}\left[\frac{\sqrt{1-\mu^2\cos^2(\frac{\pi}{M})}-\mu\sin(\frac{\pi}{M})}{(1-\mu^2)\sqrt{1-\mu^2\cos^2(\frac{\pi}{M})}}\right]
$$

(A.4)

We can simplify further the above equation by noting that:

$$
\begin{aligned}
1-\mu^2 &= \left[\sqrt{1-\mu^2\cos^2(\frac{\pi}{M})}-\mu\sin(\frac{\pi}{M})\right]\\
&\quad \times \left[\sqrt{1-\mu^2\cos^2(\frac{\pi}{M})}+\mu\sin(\frac{\pi}{M})\right]
\end{aligned}
$$

Hence, (A.4) becomes

$$
\begin{aligned}
P_M &\approx \frac{(M-1)(1-\mu^2)}{M}\frac{1}{\left[\sqrt{1-\mu^2\cos^2(\frac{\pi}{M})}+\mu\sin(\frac{\pi}{M})\right]}\\
&\quad \times \frac{1}{\sqrt{1-\mu^2\cos^2(\frac{\pi}{M})}}\\
&\approx \frac{(M-1)(1-\mu^2)}{M}\frac{1}{\left[2\mu\sin(\frac{\pi}{M})\right]\mu\sin(\frac{\pi}{M})}\\
&= \frac{(M-1)(1-\mu^2)}{2M\mu^2\sin^2(\frac{\pi}{M})} = \frac{(M-1)}{2M\bar{\gamma}_c\sin^2(\frac{\pi}{M})}
\end{aligned}
$$

where the last equality is due to the fact that (see (A.2)):

$$
\bar{\gamma}_c = \frac{\mu^2}{1-\mu^2}
$$

Therefore, we have

$$
P_M \approx \frac{(M-1)}{2M(\log_2 M)\bar{\gamma}_b\sin^2(\frac{\pi}{M})}
$$

(A.5)

Appendix B
Proof of the Decision Metrics for Unitary DSTBCs

In this section, we derive the expression of the statistic D_j mentioned in Eq. (6.10). Then, we prove that the detector for the symbol s_j is given by: $\hat{s}_j = Arg\{\max_{s_j \in S} Re\{D_j^* s_j\}\}$. Before proceeding further, it is important to note that:

1. $\{tr(\boldsymbol{\Theta}\mathbf{A}^H\mathbf{A})\}$ is real if $\boldsymbol{\Theta}$ is a Hermitian matrix, i.e. $\boldsymbol{\Theta} = \boldsymbol{\Theta}^H$. Consequently, $Im\{tr(\boldsymbol{\Theta}\mathbf{A}^H\mathbf{A})\} = 0$.

2. $tr(\boldsymbol{\Omega}\boldsymbol{\Lambda}) = tr(\boldsymbol{\Lambda}\boldsymbol{\Omega})$ if $\boldsymbol{\Omega}$ and $\boldsymbol{\Lambda}$ are square matrices.

3. $\mathbf{Z}_t^H \mathbf{W}_{t-1}^H = \mathbf{W}_t^H$.

4. $\mathbf{Z}_t = \frac{1}{\sqrt{p}} \sum_{k=1}^p (\mathbf{X}_k s_k^R + i\mathbf{Y}_k s_k^I)$, i.e., $\mathbf{Z}_t^H = \frac{1}{\sqrt{p}} \sum_{k=1}^p (\mathbf{X}_k^H s_k^R - i\mathbf{Y}_k^H s_k^I)$.

5. $\{\mathbf{X}_k\}_{k=1}^p$ and $\{\mathbf{Y}_k\}_{k=1}^p$ satisfy:

$$\mathbf{X}_k\mathbf{X}_k^H = \mathbf{I};\ \mathbf{Y}_k\mathbf{Y}_k^H = \mathbf{I} \quad \forall k \tag{B.1}$$
$$\mathbf{X}_k\mathbf{X}_j^H = -\mathbf{X}_j\mathbf{X}_k^H;\ \mathbf{Y}_k\mathbf{Y}_j^H = -\mathbf{Y}_j\mathbf{Y}_k^H \quad \forall k \neq j \tag{B.2}$$
$$\mathbf{X}_k\mathbf{Y}_j^H = \mathbf{Y}_j\mathbf{X}_k^H;\ \quad \forall k,j \tag{B.3}$$

One has:

$$\begin{aligned}
\mathbf{R}_t^H\mathbf{R}_{t-1} &= (\mathbf{A}\mathbf{W}_{t-1}\mathbf{Z}_t + \mathbf{N}_t)^H(\mathbf{A}\mathbf{W}_{t-1} + \mathbf{N}_{t-1}) \\
&= \mathbf{Z}_t^H\mathbf{W}_{t-1}^H\mathbf{A}^H\mathbf{A}\mathbf{W}_{t-1} + \mathbf{Z}_t^H\mathbf{W}_{t-1}^H\mathbf{A}^H\mathbf{N}_{t-1} \\
&+ \mathbf{N}_t^H\mathbf{A}\mathbf{W}_{t-1} + \mathbf{N}_t^H\mathbf{N}_{t-1} \tag{B.4}
\end{aligned}$$

If the noise variance is small enough, the term $\mathbf{N}_t^H \mathbf{N}_{t-1}$ is negligible. From (B.1), (B.2), (B.4) and the 2nd, 3rd, 4th notes as mentioned above, we have the following transforms:

$$
\begin{aligned}
D_j^R \;\triangleq\;& Re\{tr(\mathbf{R}_t^H \mathbf{R}_{t-1}\mathbf{X}_j)\} \\
\approx\;& \left[Re\{tr(\frac{1}{\sqrt{p}}\sum_{k=1}^{p}\mathbf{X}_k^H\mathbf{W}_{t-1}^H\mathbf{A}^H\mathbf{A}\mathbf{W}_{t-1}\mathbf{X}_j s_k^R)\} \right. \\
& \left. -\; Re\{tr(\frac{1}{\sqrt{p}}\sum_{k=1}^{p}i\mathbf{Y}_k^H\mathbf{W}_{t-1}^H\mathbf{A}^H\mathbf{A}\mathbf{W}_{t-1}\mathbf{X}_j s_k^I)\} \right] \\
& +\; Re\{tr(\mathbf{Z}_t^H\mathbf{W}_{t-1}^H\mathbf{A}^H\mathbf{N}_{t-1}\mathbf{X}_j)\} + Re\{tr(\mathbf{N}_t^H\mathbf{A}\mathbf{W}_{t-1}\mathbf{X}_j)\} \\
=\;& \left[Re\{tr(\frac{1}{\sqrt{p}}\mathbf{X}_j^H\mathbf{W}_{t-1}^H\mathbf{A}^H\mathbf{A}\mathbf{W}_{t-1}\mathbf{X}_j)s_j^R\} \right. \\
& +\; Re\{tr(\frac{1}{\sqrt{p}}\sum_{k=1,k\neq j}^{p}\mathbf{X}_k^H\mathbf{W}_{t-1}^H\mathbf{A}^H\mathbf{A}\mathbf{W}_{t-1}\mathbf{X}_j s_k^R)\} \\
& \left. -\; Re\{tr(\frac{1}{\sqrt{p}}\sum_{k=1}^{p}i\mathbf{Y}_k^H\mathbf{W}_{t-1}^H\mathbf{A}^H\mathbf{A}\mathbf{W}_{t-1}\mathbf{X}_j s_k^I)\} \right] \\
& +\; Re\{tr(\mathbf{Z}_t^H\mathbf{W}_{t-1}^H\mathbf{A}^H\mathbf{N}_{t-1}\mathbf{X}_j)\} + Re\{tr(\mathbf{N}_t^H\mathbf{A}\mathbf{W}_{t-1}\mathbf{X}_j)\} \\
=\;& \left[\frac{1}{\sqrt{p}}tr(\mathbf{A}^H\mathbf{A})s_j^R \right. \\
& +\; Re\{tr(\frac{1}{\sqrt{p}}\sum_{k=1,k\neq j}^{p}\mathbf{X}_k^H\mathbf{W}_{t-1}^H\mathbf{A}^H\mathbf{A}\mathbf{W}_{t-1}\mathbf{X}_j s_k^R)\} \\
& \left. -\; Im\{tr(\frac{1}{\sqrt{p}}\sum_{k=1}^{p}\mathbf{X}_j\mathbf{Y}_k^H s_k^I\mathbf{A}^H\mathbf{A})\} \right] + Re\{tr(\mathbf{W}_t^H\mathbf{A}^H\mathbf{N}_{t-1}\mathbf{X}_j)\} \\
& +\; Re\{tr(\mathbf{N}_t^H\mathbf{A}\mathbf{W}_{t-1}\mathbf{X}_j)\}
\end{aligned}
$$

Let $\boldsymbol{\Gamma} = \sum_{k=1}^{p}\mathbf{X}_j\mathbf{Y}_k^H s_k^I$. From (B.3), clearly, $\boldsymbol{\Gamma} = \boldsymbol{\Gamma}^H$, i.e. $\boldsymbol{\Gamma}$ is a Hermitian matrix. Therefore: $Im\{tr(\frac{1}{\sqrt{p}}\sum_{k=1}^{p}\mathbf{X}_j\mathbf{Y}_k^H s_k^I\mathbf{A}^H\mathbf{A})\}=0$. Additionally, if $\{\mathbf{X}_k\}_{k=1}^{p}$ satisfy (B.1) and (B.2) individually, then $Re\{tr(\sum_{k=1,k\neq j}^{p}\mathbf{X}_k^H\mathbf{W}_{t-1}^H\mathbf{A}^H\mathbf{A}\mathbf{W}_{t-1}\mathbf{X}_j s_k^R)\} = 0$. Hence:

$$
\begin{aligned}
D_j^R \;\approx\;& \frac{1}{\sqrt{p}}tr(\mathbf{A}^H\mathbf{A})s_j^R + Re\{tr(\mathbf{W}_t^H\mathbf{A}^H\mathbf{N}_{t-1}\mathbf{X}_j)\} \\
& + Re\{tr(\mathbf{N}_t^H\mathbf{A}\mathbf{W}_{t-1}\mathbf{X}_j)\}
\end{aligned}
$$

Similarly, we have:

$$
\begin{aligned}
D_j^I &\triangleq Re\{tr(\mathbf{R}_t^H \mathbf{R}_{t-1} i \mathbf{Y}_j)\} \\
&\approx \frac{1}{\sqrt{p}} tr(\mathbf{A}^H \mathbf{A}) s_j^I + Re\{tr(\mathbf{W}_t^H \mathbf{A}^H \mathbf{N}_{t-1} i \mathbf{Y}_j)\} \\
&+ Re\{tr(\mathbf{N}_t^H \mathbf{A} \mathbf{W}_{t-1} i \mathbf{Y}_j)\}
\end{aligned}
$$

The statistic for decoding the symbol s_j is given below:

$$
\begin{aligned}
D_j &= D_j^R + i D_j^I \\
&= \frac{1}{\sqrt{p}} tr(\mathbf{A}^H \mathbf{A}) s_j + Re\{tr(\mathbf{W}_t^H \mathbf{A}^H \mathbf{N}_{t-1} \mathbf{X}_j)\} \\
&+ Re\{tr(\mathbf{N}_t^H \mathbf{A} \mathbf{W}_{t-1} \mathbf{X}_j) + i Im\{tr(\mathbf{W}_t^H \mathbf{A}^H \mathbf{N}_{t-1} \mathbf{Y}_j)\} \\
&+ i Im\{tr(\mathbf{N}_t^H \mathbf{A} \mathbf{W}_{t-1} \mathbf{Y}_j)\} \tag{B.5}
\end{aligned}
$$

Decoding the symbol s_j is equivalent to minimizing the following expression (note that $|s_j|^2 = 1$):

$$
\begin{aligned}
|D_j - \frac{1}{\sqrt{p}} tr(\mathbf{A}^H \mathbf{A}) s_j|^2 &= D_j^* D_j + \frac{1}{p}(tr(\mathbf{A}^H \mathbf{A}))^2 \\
&- \frac{2}{\sqrt{p}} tr(\mathbf{A}^H \mathbf{A}) Re\{D_j^* s_j\}
\end{aligned}
$$

Therefore, the detector of the symbol s_j is:

$$
\hat{s}_j = Arg\{\max_{s_j \in S} Re\{D_j^* s_j\}\} \tag{B.6}
$$

The expressions (B.5), (B.6) are the aim of the proof. ∎

References

3GPP (2000). Physical channels and mapping of transport channels onto physical channels (FDD) (Release 1999). Technical Report 3GPP (Third Genenration Partnership Project), TS 25.211 V3.5.0.

3GPP (2002). STTD with adaptive transmitted power allocation. Technical Report 3GPP (Third Genenration Partnership Project), TSG RAN WG1 #26 R1-02-0711.

A.-Meraim, K., Damen, M. O., and Belfiore, J.-C. (2002). Diagonal algebraic space-time block codes. *IEEE Trans. Infor. Theory*, 48(3):628 − 636.

Adams, J. F., Lax, P. D., and Phillips, R.S. (1965). On matrices whose real linear combinations are nonsingular. *Proc. Amer. Math. Soc.*, 16:318–322.

Adams, J. F., Lax, P. D., and Phillips, R.S. (1966). Corrections to 'On matrices whose real linear combinations are nonsingular'. *Proc. Amer. Math. Soc.*, 17:945–947.

Adeli, H. and Soegiarso, R. (1999). *High-performance computing in structural engineering*. Boca CRC Press, Boca Raton.

Al-Dhahir, N. (2002). Overview and comparison of equalization schemes for space-time-coded signals with application to EDGE. *IEEE Trans. Signal Process.*, 50(10):2477 − 2488.

Alamouti, S. M. (1998). A simple transmit diversity technique for wireless communications. *IEEE J. Select. Areas Commun.*, 16(8):1451 − 1458.

Barrett, M. and Arnott, R. (1994). Adaptive antennas for mobile communications. *IEE Electronics & Communication Engineering Journal*, 6(4):203 − 214.

Beaulieu, N. C. (1999). Generation of correlated Rayleigh fading envelopes. *IEEE Commun. Lett.*, 3(6):172–174.

Beaulieu, N. C. and Merani, M. L. (2000). Efficient simulation of correlated diversity channels. *Proc. IEEE Conference on Wireless Communications and Networking WCNC 2000*, 1:207–210.

Blogh, J. S. and Hanzo, L. (2002). *Third-generation systems and intelligent wireless networking smart antennas and adaptive modulation*. John Wiley & Sons, Boca Raton.

Chen, Z. (2004). Asymptotic performance of transmit antenna selection with maximal-ratio combining for generalized selection criterion. *IEEE Commun. Lett.*, 8(4):247–249.

Chen, Z., Vucetic, B., and Yuan, J. (2003a). Comparison of three closed-loop transmit diversity schemes. *Proc. 57th IEEE Veh. Technol. Conf. VTC 2003-Spring*, 1:751–754.

Chen, Z., Vucetic, B., Yuan, J., and Lo, K. L. (2003b). Analysis of transmit antenna selection/ maximal-ratio combining in Rayleigh fading channels. *Proc. Int. Conf. Commun. Technol. ICCT 2003*, 2:1532 – 1536.

Chen, Z., Yuan, J., Vucetic, B., and Zhou, Z. (2003c). Performance of Alamouti scheme with transmit antenna selection. *Electronics Lett.*, 39(23):1666–1668.

Chen, Z., Zhu, G., Shen, J., and Liu, Y. (2003d). Differential space-time block codes from amicable orthogonal designs. *Proc. IEEE Conference on Wireless Communications and Networking WCNC 2003*, 2:768–772.

Choi, J., Choi, H. K., and Lee, H. W. (2002). An adaptive technique for transmit antenna diversity with feedback. *IEEE Trans. Veh. Technol.*, 51(4):617 – 623.

Clarke, R. H. (1968). A statistical theory of mobile-radio reception. *Bell system technical journal*, 47:957–1000.

Damen, M. O., Tewfik, A., and Belflore, J. C. (2002). A construction of a space-time code based on number theory. *IEEE Trans. Infor. Theory*, 48(3):753 – 760.

de Leon, C. A. G. D., Bean, M. C., and Garcia, J. S. (2004a). Generation of correlated Rayleigh-fading envelopes for simulating the variant behavior of indoor radio propagation channels. *Proc. 60th IEEE Veh. Technol. Conf. VTC 2004 - Fall*, 6:4245 – 4249.

de Leon, C. A. G. D., Bean, M. C., and Garcia, J. S. (2004b). On the generation of correlated Rayleigh envelopes for representing the variant behavior of the indoor radio propagation channel. *Proc. 15th IEEE Int. Symp. Personal, Indoor and Mobile Radio Commun. PIMRC 2004*, 4:2757–2761.

Electronics, LG (2002). Adaptive STTD enhancement. Technical Report 3GPP (Third Genenration Partnership Project), TSG RAN WG1# 28 R1-02-1137, WA, USA.

Electronics, LG (2004). More results on D-TxAA for MIMO. Technical Report 3GPP (Third Genenration Partnership Project), TSG RAN WG1# 37 R1-040484, Montreal, Canada.

Ericsson (2004). Text proposal for MIMO TR25.876 v1.2.0 (S-PARC). Technical Report 3GPP (Third Genenration Partnership Project), TSG RAN WG1 #36 R1-040308, Malaga, Spain.

Ertel, R. B. and Reed, J. H. (1998). Generation of two equal power correlated Rayleigh fading envelopes. *IEEE Commun. Lett.*, 2(10):276–278.

Foschini, G. J. (1996). Layered space-time architecture for wireless communication in a fading environment when using multi-element antennas. *Bell Labs Technical Journal*, 1(2):41–59.

Foschini, G. J. and Gans, M. J. (1998). *On limits of wireless communications in a fading environment when using multiple antennas*, volume 6 of *3*. Wireless Personal Communications, Printed in the Netherlands.

Fragouli, C., Krikidis, I., Levisianos, A., Polydoros, A., Dallas, P., Valkanas, A., Gallo, A. S., Vitetta, G., Matinmikko, M., and Mammela, A. (2002). State-of-the-art space-time codes. *Microsoft Internet Information Server* [Online]. Available: http://w3dpdext1.intranet.gr/stingray_docs/stingray_2D1_public.pdf [Accessed: 16 June, 2004].

Ganesan, G. (2002). *Designing space-time codes using orthogonal designs*. Doctoral dissertation, Uppsala University, Sweden.

Ganesan, G. and Stoica, P. (2000). Space-time diversity using orthogonal and amicable orthogonal designs. *Proc. IEEE Int. Conf. Acoust., Speech, and Signal Processing ICASSP '00*, 5:2561–2564.

Ganesan, G. and Stoica, P. (2001). Space-time block codes: a maximum SNR approach. *IEEE Trans. Inform. Theory*, 47(4):1650–1656.

Ganesan, G. and Stoica, P. (2002a). Differential detection based on space-time block codes. *Wireless Personal Communications: An International Journal*, 21(2):163–180.

Ganesan, G. and Stoica, P. (2002b). Differential modulation using space-time block codes. *IEEE Signal Process. Lett.*, 9(2):57–60.

Gans, M. J. (1972). A power spectral theory of propagation in the mobile radio environment. *IEEE. Trans. Veh. Technol.*, VT-21:27–38.

Geramita, A. V and Pullman, N. J. (1974). A theorem of Hurwitz and Radon and orthogonal projective modules. *Proc. Amer. Math. Soc.*, 42(1):51–56.

Geramita, A. V. and Seberry, J. (1979). *Orthogonal designs: quadratic forms and Hadamard matrices*, volume 43. Lecture notes in pure and applied mathematics, Marcel Dekker, New York and Basel.

Godara, L. C. (1997). Application of antenna arrays to mobile communications. II. Beam-forming and direction-of-arrival considerations. *Proceedings of the IEEE*, 85(8):1195 – 1245.

Gore, D. A. and Paulraj, A. J. (2001). Space-time block coding with optimal antenna selection. *Proc. IEEE Int. Conf. Acoust., Speech and Signal Processing ICASSP '01*, 4:2441–2444.

Gore, D. A. and Paulraj, A. J. (2002). MIMO antenna subset selection with space-time coding. *IEEE Trans. Signal Process.*, 50(10):2580–2588.

Griffith, E. (2004). 802.11n: The battle begins. *Enterprise ITplanet.com Networking* [Online]. Available: http://www.enterpriseitplanet.com/networking/news/article.php/3380511 [Accessed: 16 Jan., 2005].

Group, Farpoint (2003). Advanced wireless technologies: MIMO comes of age. *Airgo Network* [Online]. Available: http://www.airgonetworks.com/product_casestudies.html [Accessed: 30 Jan., 2005].

Hamalainen, J. and Wichman, R. (2002). Performance analysis of closed-loop transmit diversity in the presence of feedback errors. *Proc. 13th IEEE Int. Symp. Personal, Indoor and Mobile Radio Commun. PIMRC 2002*, 5:2297 – 2301.

Hansson, A. and Aulin, T. (1999). Generation of N correlated Rayleigh fading processes for the simulation of space-time-selective radio channels. *Proc.*

European Wireless '99 / 4th ITG Conference on Mobile Communications, pages 269 – 272.

Hassibi, B. and Hochwald, B. (2001). High-rate linear space-time codes. *Proc. IEEE Int. Conf. Acoust., Speech, and Signal Processing ICASSP '01*, 4:2461–2464.

Hassibi, B. and Hochwald, B. M. (2002). High-rate codes that are linear in space and time. *IEEE Trans. Inform. Theory*, 48(7):1804–1824.

Hochwald, B. M. and Marzetta, T. L. (2000). Unitary space-time modulation for multiple-antenna communications in Rayleigh flat fading. *IEEE Trans. Inform. Theory*, 46(2):543–564.

Hochwald, B. M. and Sweldens, W. (2000). Differential unitary space-time modulation. *IEEE Trans. Commun.*, 48(12):2041–2052.

Hughes, B. L. (2000). Differential space-time modulation. *IEEE Trans. Inform. Theory*, 46(7):2567–2578.

Jafarkhani, H. (2001). A quasi-orthogonal space-time block codes. *IEEE Trans. Commun.*, 49(1):1 – 4.

Jakes, W. C. (1974). *Microwave mobile communications*. John Wiley & Sons, New York.

Jones, V. K., Raleigh, G., and Nee, R. V. (2003). MIMO answers high-rate WLAN call. *EE Times [Online]. Available:http://eetimes.com/showArticle.jhtml?articleID=18310191 [Accessed: 16 Jan., 2005]*.

Jozefiak, T. (1976). Realization of Hurwitz-Radon matrices. *Queen's Papers on Pure and Applied Mathematics*, 36:346–351.

Katz, M., Tiirola, E., and Ylitalo, J. (2001). Combining space-time block coding with diversity antenna selection for improved downlink performance. *Proc. 54th IEEE Veh. Technol. Conf. VTC 2001 - Fall*, 1:178–182.

Katz, M. and Ylitalo, J. (2000). Extension of space-time coding to beamforming WCDMA base stations. *Proc. 51st IEEE Veh. Technol. Conf. VTC 2000 - Spring*, 2:1230 –1234.

Kharaghani, H. (2000). Arrays for orthogonal designs. *J. Combin. Designs*, 8:166 – 173.

Larsson, E. G., Ganesan, G., and Stoica, P. (2003). *Space-time block coding for wireless communications*. Cambridge University Press.

Liang, X.-B. (2003). Orthogonal designs with maximal rates. *IEEE Trans. Inform. Theory*, 49(10):2468–2503.

Liang, X.-B. and Xia, X.-G. (2003). On the nonexistence of rate-one generalized complex orthogonal designs. *IEEE Trans. Inform. Theory*, 49(11):2984 – 2988.

Liew, T. H. and Hanzo, L. (2002). Space-time codes and concatenated channel codes for wireless communications. *Proceedings of the IEEE*, 90(2):187 – 219.

Lo, T. and Tarokh, V. (1999). Space-time block coding - from a physical perspective. *Proc. IEEE Conference on Wireless Communications and Networking WCNC 1999*, 1:150–153.

Marzetta, T. L. and Hochwald, B. M. (1999). Capacity of a mobile multiple-antenna communication link in Rayleigh flat fading. *IEEE Trans. Inform. Theory*, 45(1):139–157.

Natarajan, B., Nassar, C. R., and Chandrasekhar, V. (2000). Generation of correlated Rayleigh fading envelopes for spread spectrum applications. *IEEE Commun. Lett.*, 4(1):9–11.

Nokia (2002). Text proposal of EB analysis for Tx-diversity TR (TR25.869). Technical Report 3GPP (Third Genenration Partnership Project), TSG RAN WG1 R1#24 0271.

Nokia (2004). Closed loop MIMO with 4 Tx and 2 Rx antennas. Technical Report 3GPP (Third Genenration Partnership Project), TSG RAN WG1#36 R1-040206, Malaga, Spain.

Proakis, J. G. (2001). *Digital communications*. McGraw-Hill, Boston, fourth edition.

Rappaport, T. S. (2002). *Wireless communication: principles and practice*. Prentice Hall PTR, Upper Saddle River, N.J.; London, second edition.

Rappaport, T. S., Annamalai, A., Buehrer, R. M, and Tranter, W. H. (2002). Wireless communications: past events and a future perspective. *IEEE Commun. Mag.*, 50th anniversary commemorative issue:148–161.

Salz, J. and Winters, J. H. (1994). Effect of fading correlation on adaptive arrays in digital mobile radio. *IEEE Trans. Veh. Technol.*, 43(4):1049–1057.

Samsung and SNU (2004). Text proposal for MIMO TR25.876 v1.3.0. Technical Report 3GPP (Third Genenration Partnership Project), TSG RAN WG1 #37 R1-040419, Montreal, Canada.

Sandhu, S. and Paulraj, A. (2000). Space-time block codes: a capacity perspective. *IEEE Commun. Lett.*, 4(12):384 – 386.

Seberry, J., Tran, L. C., Wang, Y., Wysocki, B. J., Wysocki, T. A., Xia, T., and Zhao, Y. (2004). New complex orthogonal space-time block codes of order eight. In T. A. Wysocki, B. Honary and Wysocki, B. J., editors, *Signal Processing for Telecommunications and Multimedia*, volume 27 of *Multimedia systems and applications*, pages 173–182. Springer, New York.

Seshadri, N. and Winters, J. H. (1993). Two signaling schemes for improving the error performance of frequency-division-duplex (FDD) transmission systems using transmitter antenna diversity. *Proc. 43rd IEEE Veh. Technol. Conf.*, pages 508 – 511.

Smith, J. I. (1975). A computer generated multipath fading simulation for mobile radio. *IEEE Trans. Veh. Technol.*, VT-24(3):39–40.

Song, A. and Xia, X.-G. (2003). Decision feedback differential detection for differential orthogonal space-time modulation with ASPK signals over frequency-non-selective fading channels. *Proc. IEEE Int. Conf. Commun. ICC'03*, 2:1253–1257.

Sorooshyari, S. and Daut, D. G. (2003). Generation of correlated Rayleigh fading envelopes for accurate performance analysis of diversity systems. *Proc. 14th IEEE Int. Symp. Personal, Indoor and Mobile Radio Commun. PIMRC 2003*, 2:1800–1804.

Standards Association, IEEE (1999). Part 11: Wireless LAN medium access control (MAC) and physical layer (PHY) specifications - High-speed physical layer in the 5 GHz band. *IEEE Standards Association [Online]. Available: http://standards.ieee.org/getieee802/download/802.11a-1999.pdf [Accessed: 12 Apr., 2005]*.

Street, D. J. (1981). *Cyclotomy and designs*. Doctoral disseration, University of Sydney, Australia.

Street, D. J. (1982). Amicable orthogonal designs of order eight. *Journal of Australian Mathematical Society, Series A*, 33:23–29.

Su, W. and Xia, X.-C. (2001). Two generalized complex orthogonal space-time block codes of rates 7/11 and 3/5 for 5 and 6 transmit antennas. *Proc. Int. Symp. Inform. Theory ISIT'01*.

Su, W. and Xia, X.-G. (2002a). A design of quasi-orthogonal space-time block codes with full diversity. *Conference Record of the 36th Asilomar Conf. Signals, Syst., Comput.*, 2:1112 – 1116.

Su, W. and Xia, X.-G. (2002b). Quasi-orthogonal space-time block codes with full diversity. *Proc. IEEE Global Telecommunications Conference GLOBE-COM '02*, 2:1098 – 1102.

Su, W. and Xia, X.-G. (2003). On space-time block codes from complex orthogonal designs. *Wireless Personal Communications*, 25(1):1–26.

Tarokh, V. and Jafarkhani, H. (2000). A differential detection scheme for transmit diversity. *IEEE J. Select. Areas Commun.*, 18(7):1169–1174.

Tarokh, V., Jafarkhani, H., and Calderbank, A. R. (1999a). Space-time block coding for wireless communications: performance results. *IEEE J. Select. Areas Commun.*, 17(3):451 – 460.

Tarokh, V., Jafarkhani, H., and Calderbank, A. R. (1999b). Space-time blocks codes from orthogonal designs. *IEEE Trans. Inform. Theory*, 45(5):1456 – 1467.

Tarokh, V., Naguib, A., Seshadri, N., and Calderbank, A. R. (1999c). Space-time codes for high data rate wireless communication: performance criteria in the presence of channel estimation errors, mobility, and multiple paths. *IEEE Trans. Commun.*, 47(2):199 – 207.

Tarokh, V., Seshadri, N., and Calderbank, A. R. (1998). Space-time codes for high data wireless communications: performance criterion and code construction. *IEEE Trans. Inform. Theory*, 44(2):744 – 765.

Telatar, I. E. (1999). Capacity of multi-antenna Gaussian channels. *European Trans. Telecom.*, 10(6):585–595.

Tirkkonen, O. and Hottinen, A. (2001). Tradeoffs between rate, puncturing and orthogonality in space-time block codes. *Proc. IEEE Int. Conf. Commun. ICC 2001*, 4:1117 – 1121.

Tirkkonen, O. and Hottinen, A. (2002). Square-matrix embeddable space-time blocks codes for complex signal constellations. *IEEE Trans. Inform. Theory*, 48(2):384 – 395.

Tran, L. C., Seberry, J., Wysocki, B. J., Wysocki, T. A., Xia, T., and Zhao, Y. (2004a). Two new complex orthogonal space-time codes for 8 transmit antennas. *IEE Electronics Lett.*, 40(1):55–56.

Tran, L. C. and Wysocki, T. A. (2003). Antenna selection scheme for wireless channels utilizing differential space-time modulation. *Proc. 7th International Symposium on Digital Signal Processing and Communications Systems DSPCS'2003*, pages 452–457.

Tran, L. C. and Wysocki, T. A. (2004). On some antenna selection techniques for wireless channels utilizing differential space-time modulation. *Proc. IEEE Conference on Wireless Communications and Networking WCNC 2004*, 2:1210–1215.

Tran, L. C., Wysocki, T. A., and Mertins, A. (2003). Improved antenna selection technique to enhance the performance of wireless communications channels. *Proc. 7th International Symposium on Signal Processing and Its Applications ISSPA 2003*, 1:257–260.

Tran, L. C., Wysocki, T. A., and Mertins, A. (2005a). A generalized algorithm for the generation of correlated Rayleigh fading envelopes in wireless channels. *To appear in the EURASIP Journal on Wireless Communications and Networking (EURASIP JWCN)*.

Tran, L. C., Wysocki, T. A., Mertins, A., and Seberry, J. (2004b). Transmitter diversity antenna selection techniques for wireless channels utilizing differential space-time block codes. *Submitted to J. Telecom. Inform. Technol. (JTIT)*.

Tran, L. C., Wysocki, T. A., Mertins, A., and Seberry, J. (2005b). A generalized algorithm for the generation of correlated Rayleigh fading envelopes. *Proc. 6th IEEE International Symposium on a World of Wireless, Mobile and Multimedia Networks WOWMOM 2005*, pages 213–218.

Tran, L. C., Wysocki, T. A., Mertins, A., and Seberry, J. (2005c). A generalized algorithm for the generation of correlated Rayleigh fading envelopes in radio channels. *Proc. 19th IEEE International Parallel and Distributed Processing Symposium IPDPS'05*, pages 238b – 238b.

Tran, L. C., Wysocki, T. A., Seberry, J., and Mertins, A. (2004c). The effect of imperfect carrier recovery on the performance of the diversity antenna selection technique in wireless channels utilizing DSTM. *Proc. 2nd IEEE Int. Conf. Inform. Technol.: Research and Education ITRE 2004*, 1:15–18.

Tran, L. C., Wysocki, T. A., Seberry, J., and Mertins, A. (2004d). Multi-modulation schemes to increase the rate of space-time block codes in Rayleigh fading channels. *Proc. 2nd IEEE Int. Conf. Inform. Technol.: Research and Education ITRE 2004*, 1:19–23.

Tran, L. C., Wysocki, T. A., Seberry, J., and Mertins, A. (2004e). On multi-modulation schemes to increase the rate of space-time block codes. *Proc. 12th European Signal Processing Conference EUSIPCO 2004*, 3:1841–1844.

Tran, L. C., Wysocki, T. A., Seberry, J., Mertins, A., and Spence, Sarah A. (2004f). Generalized Williamson and Wallis-Whiteman constructions for improved square complex orthogonal space-time block codes. *Submitted to IEEE Trans. Inform. Theory*.

Tran, L. C., Wysocki, T. A., Seberry, J., Mertins, A., and Spence, Sarah A. (2005d). Generalized Williamson and Wallis-Whiteman constructions for improved square order-8 CO STBCs. *To appear in Proc. 16th Annual IEEE Int. Symp. Personal, Indoor and Mobile Radio Commun. PIMRC 2005*.

Verdin, D. and Tozer, T. C. (1993). Generating a fading process for the simulation of land-mobile radio communications. *Electronics Lett.*, 29(23):2011–2012.

Vucetic, B. and Yuan, J. (2003). *Space-time coding*. John Wiley & Sons, Hoboken, NJ.

Wang, H. and Xia, X.-G (2003). Upper bounds of rates of complex orthogonal space-time block codes. *IEEE Trans. Inform. Theory*, 49(10):2788–2795.

Webb, W. T. and Hanzo, L. (1994). *Modern quadrature amplitude modulation - principles and applications for fixed and wireless communications*. Pentech Press Limited, London.

Winters, J. H. (1994). The diversity gain of transmit diversity in wireless systems with Rayleigh fading. *Proc. IEEE Int. Conf. Commun. ICC '94*, 2:1121 – 1125.

Winters, J. H. (1998). The diversity gain of transmit diversity in wireless systems with Rayleigh fading. *IEEE Trans. Veh. Technol.*, 47(1):119 – 123.

Wittneben, A. (1991). Basestation modulation diversity for digital simulcast. *Proc. 41st IEEE Veh. Technol. Conf. VTC '91*, pages 848 – 853.

Wittneben, A. (1993). A new bandwidth efficient transmit antenna modulation diversity scheme for linear digital modulation. *Proc. IEEE Int. Conf. Commun. ICC '93*, 3:1630 – 1634.

Wolfe, W. (1975). *Orthogonal designs - amicable orthogonal designs - some algebraic and combinatorial techniques*. Doctoral disseration, Queen's University, Kingston, Ontario, Canada.

Wolfe, W. (1976). Amicable orthogonal designs - existence. *Can. J. Math.*, 28(5):1006–1020.

Xiaofeng, T., Harald, H., Zhuizhuan, Y., Haiyan, Q., and Ping, Z. (2001). Closed loop space-time block code. *Proc. 54th IEEE Veh. Technol. Conf. VTC'2001-Fall*, 2:1093–1096.

Young, D. J. and Beaulieu, N. C. (1996). On the generation of correlated Rayleigh random variates by inverse discrete Fourier transform. *Proc. 5th IEEE International Conference on Universal Personal Communications ICUPC*, 1:231–253.

Young, D. J. and Beaulieu, N. C. (1998). A quantitative evaluation of generation methods for correlated Rayleigh random variates. *Proc. IEEE Global Telecommunications Conference GLOBECOM 98*, 6:3332–3337.

Young, D. J. and Beaulieu, N. C. (2000). The generation of correlated Rayleigh random variates by inverse discrete Fourier transform. *IEEE Trans. Commun.*, 48(7):1114–1127.

Yuen, C., Guan, Y. L., and Tjhung, T. T. (2004). Orthogonal space-time block code from amicable orthogonal design. *Proc. IEEE. Int. Conf. Acoustic, Speech and Signal Processing ICASSP 2004*, 4:469–472.

Zhao, Y., Seberry, J., Xia, T., Wang, Y., Wysocki, B. J., Wysocki, T. A., and Tran, L. C. (2005a). Amicable orthogonal designs of order 8 for complex space-time block codes. *To appear in The Australasian Journal of Combinatorics (AJC)*.

Zhao, Y., Wang, Y., and Seberry, J. (2005b). On amicable orthogonal designs of order 8. *To appear in The Australasian Journal of Combinatorics (AJC)*.

Index